The Study of Astro

to the capacities of youth

In twelve familiar dialogues, between a tutor and his pupil explaining the general phænomena of the heavenly bodies, the theory of the tides, &c.

teacher of astronomy John Stedman

Alpha Editions

This edition published in 2024

ISBN : 9789364734417

Design and Setting By
Alpha Editions
www.alphaedis.com
Email - info@alphaedis.com

As per information held with us this book is in Public Domain.
This book is a reproduction of an important historical work. Alpha Editions uses the best technology to reproduce historical work in the same manner it was first published to preserve its original nature. Any marks or number seen are left intentionally to preserve its true form.

Contents

PREFACE. .... - 1 -

DIALOGUE I. .... - 2 -

DIALOGUE II. .... - 6 -

DIALOGUE III. .... - 11 -

DIALOGUE IV. .... - 15 -

DIALOGUE V. .... - 19 -

DIALOGUE VI. .... - 25 -

DIALOGUE VII. .... - 30 -

DIALOGUE VIII. .... - 35 -

DIALOGUE IX. .... - 41 -

DIALOGUE X. .... - 47 -

DIALOGUE XI. .... - 55 -

DIALOGUE XII. .... - 63 -

PREFACE.

It has long been a matter of surprize to those who are interested in the education of youth, that, among the numerous publications intended for their improvement, so few attempts have been made to facilitate the study of Astronomy.

Many excellent treatises have been written on this important and useful science; but if it be considered that they abound with technical terms, unintelligible to juvenile minds, it cannot be expected that they should derive any great advantage from the perusal of them.

To remove these difficulties, the Author has endeavoured, whenever he had occasion to use them, to give such illustrations as to leave no doubt on the young student's mind respecting their true meaning.

The subject appeared to him to be best calculated for dialogues, which are certainly more agreeable as well as more perspicuous to young persons, than the discouraging formality of a treatise. And it is presumed the language will be found natural and easy.

In the order he has chosen, he has been careful not to introduce any thing new, till the former part, on which it depends, has been clearly explained.

On the whole, it has been his aim to render it as concise and plain as the nature of the subject will admit; and he flatters himself, that at a time when the sciences are so universally studied, the introduction now offered to the public will not be unacceptable.

DIALOGUE I.

Tutor.

Well, Sir! I suppose this early visit is in consequence of my promise, and your anxiety to become an astronomer.

Pupil. It is, Sir.—And as astronomy is a science of which I have a very imperfect idea, I must beg of you to explain it to me.

Tutor. That I shall do with pleasure. But you surely cannot wholly forget what I have formerly told you. However, as I mean to treat the subject as if you had no previous knowledge of it, you will have an opportunity from what you can recollect, to make such remarks, and ask such questions, as may appear most material to you.

Pupil. I thank you, Sir, it is just what I wish.

Tutor. By astronomy then is meant a knowledge of the heavenly bodies, the sun, moon, planets, comets, and stars, respecting their nature, magnitudes, distances, motions, &c.

Pupil. I fear I shall find it a difficult study.

Tutor. Have patience.——

"The wise and prudent conquer difficulties,

"By daring to attempt them. Sloth and folly

"Shiver and shrink at sight of toil and danger,

"And make the impossibility they fear."

Pupil. This gives me encouragement, and, if you will have patience with me, I will endeavour to profit by your instructions.——Pray, Sir, what is the sun?

Tutor. The sun, the source of light and heat, has been considered a globe of fire, round which seven other spherical bodies revolve at different distances from him, and in different periods of time, from west by south to east. These are the planets[1].

Pupil. Any round ball is a globe, is it not?

Tutor. A sphere or globe is defined a round solid body, every part of whose surface is equally distant from a point within called its center; and a line drawn from one side through the center to the opposite side, is called its diameter.

PUPIL. You say the sun has been considered a globe of fire. Is he not now thought to be so?

TUTOR. [2]Doctor Herschell, from some late observations, is of a different opinion.—But what think you of his magnitude?

PUPIL. I really cannot conjecture.—This I know, that when I saw him through the fog the other day, he appeared about the size of a common plate.

TUTOR. You must not always judge by appearances. You will find that there is a material difference between his real and apparent magnitude, which I think you will be convinced of when I tell you, that he is no less than 95 millions of miles from our earth.

PUPIL. Ninety-five millions of miles! You astonish me.

TUTOR. You will, I dare say, be no less surprized at being told, that he is more than a million of times as large as our earth.

PUPIL. It is almost incredible! And what are the planets?

TUTOR. The planets are opaque, that is dark bodies, which receive their light from the sun; and, as I told you, revolve about him. The first, or that nearest the sun, is called Mercury, the next Venus, then the Earth, Mars, Jupiter, Saturn, and Georgian, or the Georgium Sidus.[3] These are called primary planets.

PUPIL. Are there then any others?

TUTOR. Yes. There are fourteen others, which move round their respective primaries as their centers, and with them round the sun, and are called secondaries, satellites or moons.

PUPIL. Have all the primaries secondaries?

TUTOR. Only four of them have moons. The earth, I need not tell you, has one; Jupiter has four; Saturn seven, besides a stupendous ring which surrounds his body; and Georgian two.

PUPIL. In what time, and at what distances, from the sun, do the planets perform their periodical revolutions?

TUTOR. *Mercury* revolves about the sun in 88 days, at the distance of 36 millions of miles.

Venus, at the distance of 68 millions of miles, completes her revolution in 224 days.

Earth, on which we live, at the distance of 95 millions of miles, performs its period in one year.[4]

Mars, at the distance of 145 millions of miles, in little less than two of our years.

Jupiter, at the distance of 494 millions of miles, in near 12 years.

Saturn, at the distance of 906 millions of miles, in about 30 years.

Georgian, discovered a few years since by Dr. Herschell, performs its period at the distance of 1812 millions of miles, in about 83 years.[5]

PUPIL. What proportion does the earth bear in magnitude to the other planets?

TUTOR. The earth is fourteen times as large as Mercury, very little larger than Venus, and three times as large as Mars. But Jupiter is more than fourteen hundred times as large as the earth; Saturn above a thousand times as large, exclusive of his ring; and Georgian eighty-two times as large.

PUPIL. Have you any thing else, Sir, to remark concerning the planets?

TUTOR. There are several other things I intend to make you acquainted with, namely, their nature, appearances, motions, &c. At present I shall only say, that Mercury and Venus are called [6]inferior planets, their orbits or paths described in going round the sun, being within that of the earth; and the other four, whose orbits are without the earth's orbit, [7]superior planets.

PUPIL. There is one thing more I wish to know, if——

TUTOR. I suppose you were going to say if not too much trouble; that is quite unnecessary, as you well know that where I see a desire to learn, teaching is to me a pleasure.—What is it?

PUPIL. That you will be so kind as to inform me what the comets are, and if they have any motion?

TUTOR. The knowledge we have of comets is very imperfect, as they afford few observations on which to ground conjecture. They are generally supposed to be planetary bodies, forming a part of our system: for, like the planets, they revolve about the sun, but in different directions, and in extremely long elliptic curves, being sometimes near the sun, at others staying far beyond the orbit of the outermost planet; whereas the orbits of the planets are nearly circular. The period of one, which appeared in 1680, is computed to be 575 years.

PUPIL. Whence do they derive their name?

TUTOR. From *Cometa*, a *hairy star*, because they appear with long tails, somewhat resembling hair: some, however, have been seen without this appendage, as well defined and round as planets.

PUPIL. You say *our* system: what am I to understand by it?

TUTOR. The word system, in an astronomical sense, means a number of bodies moving round one common center or point: and, because the planets and comets revolve about the sun, it is called the *Solar System* (Plate I. fig. 2.); and we say *our* system, as the earth is one of the planets. Other systems have been invented for solving the appearances and motions of the heavenly bodies, a description of which I shall leave till I next see you.

---

1. From *Planeta*, roving or wandering.

2. See his letter read at the Royal Society, December 18th, 1794.

3. Their characters are,

Sun,	Mercs.	Venus,	Earth,	Mars,	Jup.	Saturn,	Georgian,
☉	☿	♀	♁	♂	♃	♄	♅

4. The motion of the earth in its orbit is at the rate of 68 thousand miles an hour.

5. As the distances of the planets, when marked in miles, are a burthen to the memory, astronomers often express their mean distances in a shorter way, by supposing the distance of the earth from the sun to be divided into ten parts. Mercury may then be estimated at four of such parts from the sun, Venus at seven, the Earth at ten, Mars at fifteen, Jupiter at fifty-two such parts, Saturn at ninety-five, and Georgian 190 parts. See Plate I. Fig. 1.

These are calculated by multiplying the respective distances of the planets by 10, and dividing by 95, the mean distance of the earth from the sun; and may be set off by any scale of equal parts.

6. Perhaps with more propriety *interior* or *inward*.

7. *Exterior* or *outward*.

DIALOGUE II.

Pupil.

I am afraid, Sir, I am come before you are prepared for me: but the very great pleasure I received yesterday, induced me to be with you as early as possible.

Tutor. I am glad to see you, and happy to find you are so well pleased with your difficult study. It will, I assure you, give you more exalted ideas of the Deity than any that I know of. The Psalmist was undoubtedly of this opinion when he said, The Heavens declare the glory of God, and the Firmament sheweth his handy work.

Pupil. I will no longer call it a difficult, but a pleasing study, and feel myself ashamed at having used the expression. I shall now beg you to explain to me the different systems.

Tutor. The system I have been describing to you was known and taught by Pythagoras, a Greek philosopher, who flourished about 500 years before Christ, as he found it impossible, in any other way, to give a consistent account of the heavenly motions.

This system, however, was so extremely opposite to all the prejudices of sense and opinion, that it never made any great progress, nor was ever widely spread in the ancient world.

Ptolemy, an Egyptian philosopher, who flourished 130 years after Christ, supposed that the earth was fixed in the center, and that the sun and the rest of the heavenly bodies moved round it in twenty-four hours, or one natural day, as this seemed to correspond with the sensible appearances of the cœlestial motions. This system was maintained from the time of Ptolemy to the revival of learning in the sixteenth century.

At length, Copernicus, a native of Poland, a bold and original genius, adopted the Pythagorean system, and published it to the world in the year 1530. This doctrine had been so long in obscurity, that the restorer of it was considered as the inventor.

Europe, however, was still immersed in ignorance; and the general ideas of the world were not able to keep pace with those of a refined philosophy. This occasioned Copernicus to have few abettors, but many opponents. Tycho Brahe, in particular, a noble Dane, sensible of the defects of the Ptolemaic system, but unwilling to acknowledge the motion of the earth, endeavoured, about 1586, to establish a new system of his own; but, as this proved to be still more absurd than that of Ptolemy, it was soon exploded, and gave way to the [8]Copernican or true Solar System.

PUPIL. I confess, I should have thought with Ptolemy, that the earth was in the center, and that the sun moved round it.

TUTOR. You must at present content yourself with knowing that it is not so; and it shall be my business to prove it.

PUPIL. May I beg the favour of the information you intended respecting the planets?

TUTOR. I will grant it with pleasure. The planets are spherical bodies, which appear like stars, but are not luminous; that is, they have no light in themselves; though they give us light; for they shine by reflecting the light of the sun.

PUPIL. You say, Sir, that they appear like stars; if so, how am I to know them from stars?

TUTOR. Very easily: for the stars, or as they are more properly called fixed stars, always keep the same situation with respect to each other; whereas the planets, as they move round the sun, must be continually changing their places among the fixed stars, and with one another.

PUPIL. Is there any other method of distinguishing them besides what you have mentioned?

TUTOR. Yes. The planets never twinkle like the fixed stars, and are seen earliest in the evening and latest in the morning.

PUPIL. How is the twinkling of the stars in a clear night accounted for?

TUTOR. It arises from the continual agitation of the air or atmosphere through which we view them; the particles of air being always in motion, will cause a twinkling in any distant luminous body, which shines with a strong light.

PUPIL. Then, I suppose, the planets not being luminous, is the reason why they do not twinkle.

TUTOR. Most certainly. The feeble light with which they shine is not sufficient to cause such an appearance.

PUPIL. Have the stars then light in themselves?

TUTOR. They undoubtedly shine with their own native light, or we should not see even the nearest of them: the distance being so immensely great, that if a cannon-ball were to travel from it to the sun, with the same velocity with which it left the cannon, it would be more than 1 million, 868 thousand years, before it reached it.[9]

PUPIL. This is wonderful indeed! what then are they supposed to be?

TUTOR. Suns.

PUPIL. Suns! the fixed stars suns!

TUTOR. Yes, suns.

"One sun by day, by night ten thousand shine."

And what will increase your astonishment, each of them is the center of a system of planets, which move round him.[10]

"Observe how system into system runs."

"What other planets circle other suns."

PUPIL. I am almost lost.—I used to think they were designed to give us light.

TUTOR. This is a vulgar error.—They were doubtless created for a much nobler purpose, since thousands of them are invisible to us without the help of a telescope; and we receive more light from the moon than from all the stars together.

PUPIL. How do you know they are suns? Is their being luminous a proof of their being so?

TUTOR. No. But we know that the sun shines with his own light on all the planets belonging to our system; and from what I have told you, have the greatest reason to believe that the stars shine with their own light: we therefore from analogy conclude, that they are so many suns conveying light and heat to other worlds[11].

PUPIL. Are there then other worlds besides this we live in?

TUTOR. Consider.—Has not the earth we inhabit a moon to enlighten it?

PUPIL. Yes, Sir.

TUTOR. And have I not told you that Jupiter, Saturn, and Georgian, have also moons?

PUPIL. This I well remember.

TUTOR. For what purpose then do you suppose those orbs were designed?

PUPIL. Indeed, I cannot tell.

TUTOR. You surely cannot imagine that they were intended for our use, since we knew nothing of them till after the invention of telescopes.

PUPIL. That is what I think no one can suppose.

TUTOR. And do not all the planets enjoy the benefit of the sun in common with us?

PUPIL. Undoubtedly.

TUTOR. Well, then; of what use would the light and heat be which is conveyed to them from the sun; or the light which they receive from their moons if there are no inhabitants?

PUPIL. I know of none.

TUTOR. Can you then have any doubt about their being inhabited?

PUPIL. No, Sir.—But you say that the stars are suns, each of which is the center of a system of planets or worlds.

TUTOR. If you are satisfied that the planets belonging to our system are inhabited, and that the fixed stars are suns, the centers of other systems, what reasonable objection can you have to all the planets in the universe being so?

PUPIL. It is what I cannot comprehend.

TUTOR. It may be so.—But is not the same Almighty Power, who does nothing in vain, as capable of making ten thousand worlds if he pleased, as well as one?

PUPIL. I will not presume to dispute his power; but are we not told that all mankind descended from Adam?

TUTOR. Yes; Moses wrote concerning this earth, he has not made us acquainted with the inhabitants of the other planets: for aught we know they might descend from other Adams.—To-morrow evening, I hope to see you again.

---

8. See Plate I. fig. 2.

9. The distance of Syrius is 18,717,442,690,526 miles. A cannon-ball going at the rate of 1143 miles an hour, would only reach the sun in about 1,868,307 years, 88 days.

Adams's Lectures, vol. 4. page 44.

10. Dr. Herschell says, that in some clusters of stars he has observed, they appear too close together to admit any planets to revolve about them.

11. Dr. Herschell thinks it probable that the sun and fixed stars may be inhabited.

DIALOGUE III.

Pupil.

I recollect, Sir, you mentioned last night, that the planets appear like stars. Our earth is a planet; how can it have the appearance of a star?

Tutor. If you were on the planet Venus, the earth would have as much the appearance of a star as Venus has to us.

Pupil. But Venus appears amongst the fixed stars.

Tutor. Yes. And so would the earth appear from Venus.

Pupil. How can it be?

Tutor. Because, in whatever part of the universe we are, we appear to be in the center of a concave, that is hollow, sphere, where remote objects appear at equal distances from us: so that, whether we are on the planet Venus or on the earth, in this particular the effect will be the same.

Pupil. Then the light *we* receive from the sun is by reflection conveyed to the other planets.

Tutor. No doubt of it. And our earth appears as a moon to the inhabitants of the moon, and undergoes the various changes of that planet.

Pupil. Have you any proof of this, Sir?

Tutor. Nothing can be clearer; for, on a fine evening, soon after the change of the moon, when the earth appears nearly as a full moon to the moon, and we see a faint streak of light, the whole body of the moon is visible to us.

Pupil. I remember to have seen it.

Tutor. You do?—The earth then will appear there thirteen times as large as the moon does to us; of course it must reflect a strong light on the body of the moon, and it is by that light we see that part of the moon which is turned from the sun.

Pupil. Is the earth, then, only thirteen times as big as the moon?

Tutor. In solidity it is about fifty times as large; but its disc or face is only thirteen times.

Pupil. What is the moon's distance from the earth?

TUTOR. 240 thousand miles, which is about 400 times less than that of the sun.

PUPIL. And yet she appears as far distant as the sun.

TUTOR. You are now, I hope, convinced of what I said relative to distant objects.

PUPIL. I am, Sir: and I suppose the reason of the moon's appearing as large as the sun, is because she is so much nearer to us.

TUTOR. It is so.—For, at a total eclipse of the sun, which happens when the moon is in a right line between the sun and the earth, the sun is obscured from our sight, although his disc is 160 thousand times as large as that of the moon. In like manner would the moon, when at full, be hid by placing your cricket-ball in a line between your eye and her, yet, you know, the ball is not so large as the moon; but being nearer the eye, it is apparently so.

PUPIL. This is very clear. But——

TUTOR. I conjecture you were going to ask me to explain the nature of eclipses.

PUPIL. That was certainly my intention, Sir.

TUTOR. There are other things you must be made acquainted with before you will be able to comprehend it, and which I will endeavour to make you understand before we enter on the subject.

PUPIL. Whenever you please, Sir.

TUTOR. You have taken a view of the earth from the planet Venus.—Suppose I transport you to one of the planets belonging to another system; what description do you think you should give of it?

PUPIL. I must consider. What I now call a star would be a sun. The planets of that system I should see as I now do those belonging to ours: our sun would be a star; and the earth, with all the other planets, would be invisible.

TUTOR. Very well, Sir. Can you then find it difficult to conceive that all the stars are as far from each other in unbounded space as our sun is from the nearest star?

PUPIL. It is hard to conceive: but when I consider that wherever I am, every remote object appears at an equal distance from me, the difficulty vanishes.

TUTOR. That you might form some idea of the immense distance of the fixed stars, you must recollect, I mentioned the time a cannon-ball would be in reaching the nearest of them.

PUPIL. I do, Sir. More than 1,868,000 years.

TUTOR. You have an excellent memory. I suppose then you know the distance of the earth from the sun?

PUPIL. Yes, Sir. I wrote it down; and, it made so strong an impression on my memory, that I believe I shall never forget it.—95 millions of miles.

TUTOR. Now, suppose the earth to be in that part of its orbit which is nearest to the star, it would be 95 millions of miles nearer to it than the sun is.

PUPIL. Certainly.

TUTOR. And, in the opposite side of its orbit, as much farther from the star.

PUPIL. Without doubt.

TUTOR. Then you find that the earth is 190 millions of miles nearer to the star at one time of the year than it is at another; and yet the magnitude of the star does not appear the least altered, nor is its distance affected by it.

PUPIL. A proof of its amazing distance.—I was going to ask a silly question.

TUTOR. What is it? perhaps not so simple as you may imagine.

PUPIL. Whether the most conspicuous stars are not supposed to be the nearest to us?

TUTOR. Undoubtedly.—And are called stars of the first magnitude; the next in splendor, stars of the second magnitude; and so on to the sixth magnitude; and those beyond, which are not visible to the naked eye, are called telescopic stars.

PUPIL. The distance of the telescopic stars must be great indeed, beyond all conception.

TUTOR. You judge rightly; and their numbers are beyond all computation. Doctor Herschell says, he has not a doubt but that the broad circle in the heavens, called the Milky Way, is a most extensive stratum of stars, he having discovered in it many thousands. Besides, some stars appear to him double, others treble, &c. not that they are really so, but are stars at different distances from us, which appear nearly in a right line.

"As in the milky-way a shining white

"O'erflows the heav'ns with one continued light,

"That not a single star can shew his rays,

"Whilst jointly all promote the common blaze."

PUPIL. I have heard of numbering the stars; but that, I find, is impossible.

TUTOR. If you mean that immense host of stars I have been describing, it is impossible; but, though in a clear winter's night, without moonshine, they seem to be innumerable, which is owing to their strong sparkling, and our looking at them in a confused manner; yet when the whole firmament is divided as it has been done by the ancients, the number that can be seen at a time, by the naked eye, is not above a thousand.

PUPIL. Pray, Sir, how did the ancients divide the firmament?

TUTOR. I would willingly answer your question; but, as I find I shall not have time to give you that information I wish, I shall postpone it till I see you to-morrow evening.

DIALOGUE IV.

TUTOR.

The ancients, in reducing astronomy to a science, combined the fixed stars into constellations, allowing several stars to make one constellation: and, for the better distinguishing and observing them, they reduced the constellations to the forms of animals, or to the images of some known things, by which means they were enabled to signify to others any particular star they meant to notice. Job mentions two of the constellations, namely, Orion and Pleiades, which shews the study of astronomy to be very ancient.

PUPIL. Pray, Sir, how may I know them?

TUTOR. By studying the use of the cælestial globe, on which they are drawn.

PUPIL. Will you be kind enough to instruct me, Sir?

TUTOR. At some future time I probably may: at present you are not prepared for it.

PUPIL. I am satisfied.—Have you any thing more to remark of the constellations, Sir?

TUTOR. Yes. The situation of the planets, as they are continually changing their places, could not be pointed out without first dividing the stars into constellations: hence, necessity was the mother of invention.

PUPIL. And I think a very ingenious one.—If I may be allowed a comparison, I will suppose the different kingdoms of the world on my dissected map, to represent so many constellations; then, if I hear of London, I know it is in England; if of Paris, in France; of Lisbon, in Portugal; and so on. These I would compare with stars of the first magnitude, being the chief cities of their respective kingdoms; inferior cities, stars of the second magnitude; principal towns of the third, &c.

TUTOR. A very apt comparison indeed. Now if you hear of a traveller setting off from London to Dover, thence to Calais, Paris, Bern, and so on to Rome, you know that he must go through part of England, Flanders, France, Switzerland, and Italy, passing many towns and villages on his way.

PUPIL. That is very evident.

TUTOR. Very well, then; in like manner would the planets, if seen from the sun, be traced from star to star, from constellation to constellation, through their whole periods.

PUPIL. It is not possible to view them from the sun, surely, is it?

TUTOR. No, certainly.

PUPIL. Why then do you say if seen from the sun?

TUTOR. Because it is there only their motions can appear uniform; as seen from the earth they apparently move very irregularly.—Suppose you were in the center of a circular course; and, whilst a horse was going round, you kept your eye on him: cannot you conceive that you should see him run round the course in a regular manner, moving the whole time the same way?

PUPIL. It is not at all difficult to conceive.

TUTOR. Again. Imagine yourself placed at a considerable distance on the outside of the course, where you could see the horse the whole time he was going round, would he appear to move as uniformly as before?

PUPIL. Certainly not: on the opposite side of the course his motion would be the same as when I stood in the center of it; when he was approaching me, I should scarcely see him move; in that part of the course next to me he would move in a direction contrary to what he did at first; and again when going from me, his motion would be scarcely visible.

TUTOR. This I think will give you a tolerable idea of the irregular motion of the inferior planets, as seen from the earth. When farthest from us their motion is said to be direct; when nearest to us retrograde, because they appear to be moving back again; and, when approaching, or going from us, we say they are stationary; because, if then observed in a line with any particular star, they will continue so for a considerable time: now these appearances could not happen if they moved round the earth.

PUPIL. Nothing can be plainer: for if the earth were in the center we should always see them move the same way.

TUTOR. When the planet is nearest to us, that is in a line between us and the sun, we say it is in its inferior conjunction; when farthest from us, and the sun is between us and the planet, in its superior conjunction. But the superior planets have alternately a conjunction and an opposition.

PUPIL. A conjunction, I suppose, when the sun is between the earth and the planet, and an opposition when the earth is between the sun and the planet; that is, when the planet is nearest to us, and appears to be opposite to the sun?

TUTOR. You are right.—Therefore, when in conjunction it rises and sets, nearly with the sun; but in opposition, it rises nearly when the sun sets, and sets when he rises.

PUPIL. Why do you say nearly, Sir?

TUTOR. Because it cannot be exactly, but when the sun, earth, and planet are in a *right* line, which seldom happens.

PUPIL. How do you account for this, Sir?

TUTOR. At present I fear you will not be able to comprehend what I wish to explain, as I must use a term you are unacquainted with. The reason is, that the planets are very seldom in or near their nodes at their conjunctions or oppositions.

PUPIL. I do not indeed understand what you mean by the word *nodes*.

TUTOR. It will be explained to you in due time, and I shall conclude this evening with a few more remarks relative to the appearance of the planets.

PUPIL. Any thing you please, Sir.

TUTOR. You know that the planets, being opaque bodies, receive their light from the sun; and that only that part which is turned to the sun can be enlightened by him, whilst the opposite side must remain in darkness.

PUPIL. This is self-evident: if I hold my ball to the candle it will have the same effect.

TUTOR. Tell me then how you think they will appear as seen from the earth.

PUPIL. If, when you shewed me Venus, she had not appeared perfectly round, I should say that, both before and after her superior conjunction I should see her nearly with a full face; when stationary, only half enlightened, like the moon at first quarter; because, an equal portion of the dark and bright parts will be turned towards us; the bright part will be decreasing till her inferior conjunction, when the dark side will be turned towards us, and consequently invisible; the light will then increase; and, when she is again stationary, she will appear like the moon at last quarter.

TUTOR. When seen through a telescope she has the different appearances you have mentioned; and when I next see you I will shew you that both Venus and Mercury may sometimes be seen when in their inferior conjunctions; the superior planets always appear with nearly a full face.

PUPIL. How are the planets distinguished from each other?

TUTOR. *Mercury*, from his vicinity to the sun, is seldom seen, being lost in the splendor of the solar brightness. When seen, he emits a very bright white light.

Venus, known by the names of the morning and evening star, is the brightest, and to appearance, the largest of all the planets; her light is of a white colour, and so considerable, that in a dusky place she projects a sensible shade. She is visible only for three or four hours in the morning or evening, according as she is before or after the sun.

Mars is the least bright of all the planets. He appears of a dusky reddish hue, and much larger at some periods than at others, according as he is nearer to, or farther from us.

Jupiter is distinguished by his peculiar magnitude and light. To the naked eye he appears almost as large as Venus, but not altogether so bright.

Saturn shines but with a pale feeble light, less bright than Jupiter, though less ruddy than Mars.

The Georgium Sidus cannot be readily perceived without the assistance of a telescope.

DIALOGUE V.

TUTOR.

Before I proceed to explain what I promised you, it is necessary you should be informed that the earth as seen from the sun, in its periodical revolutions, will describe a circle among the stars which astronomers call the *ecliptic*, and sometimes *the sun's annual path*, because the sun, as seen from the earth, always appears in that line.

PUPIL. Do not all the planets move in the ecliptic?

TUTOR. No.—On account of the obliquity of their orbits, they are, in every revolution, one half of their periods above the ecliptic, and the other half below it.

PUPIL. I think I comprehend your meaning; but shall be obliged to you, Sir, if you can make it clearer to me.

TUTOR. I have here a little design, (Plate II. Fig. 1.) which will answer our purpose: where S represents the sun; ABCD, the orbit of the earth; and EFGH, the orbit of one of the inferior planets, suppose Venus.

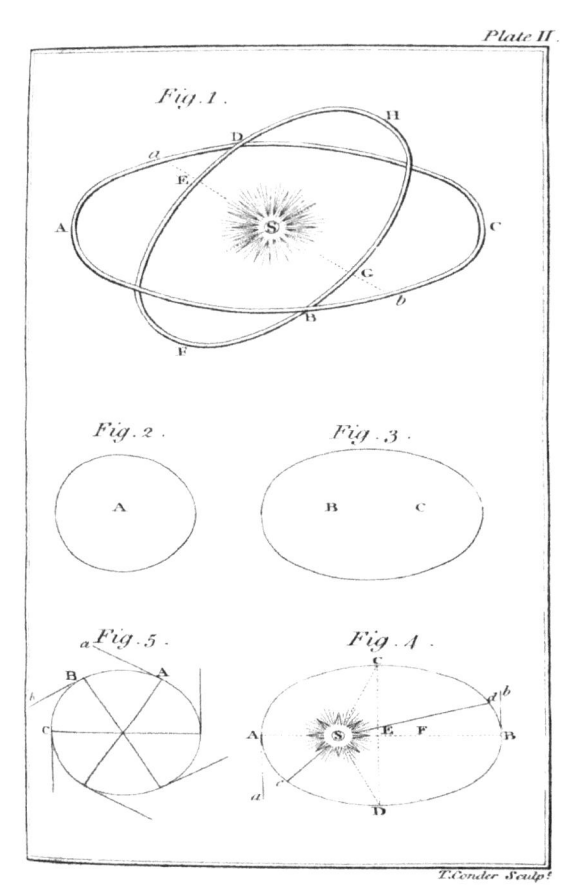

PUPIL. Now I understand it perfectly: the half EHG rises above, and the other half EFG sinks below it, from the points EG, which I perceive are in a line with the orbit of the earth. But pray, Sir, have you any name for that dotted line?

TUTOR. Yes, it is called the *line* of the nodes; and the points EG the *nodes* of the planet: the latter is called the ascending node, because, when the planet is in G, it is ascending or rising above the orbit of the earth; or, which is the same thing, above the ecliptic: and when in E, it is descending or sinking below it, whence *it* is called the descending node. But you must remember that the orbits of all the planets do not cross or intersect the ecliptic in the same points; but that their nodes or intersections are at different parts of it.

PUPIL. How can the orbit of the earth and the ecliptic be the same?

TUTOR. They are very different; but being in the same plane, if the orbit of any planet inclines to one it must incline equally to the other.

- 20 -

PUPIL. You will, I fear, Sir, think me very stupid: but I must beg of you to inform me what you mean by a plane?

TUTOR. Any flat surface is a plane. You may therefore suppose the edge of a round tea-table to represent the ecliptic, and a circle within it, drawn from the center of the table, the orbit of the earth: will they not be both in the same plane?

PUPIL. Certainly.

TUTOR. You must not imagine, when I am speaking to you of the plane of the ecliptic, or plane of the earth's orbit, that it is a visible flat surface, or, in speaking of the orbits of the planets, I mean solid rings.—No. The planets perform their revolutions with the utmost regularity, in unbounded space; and, like a bird thro' the air, leave no track behind them.

PUPIL. How then are they retained in their orbits?

TUTOR. The question, I confess, is natural, and is what I expected; but I must of necessity postpone it to another opportunity; and shall now fulfil the promise I made of shewing you in what manner the inferior planets may be seen when in their inferior conjunctions. Cast your eye again on the little design I gave you, and consider, if Venus were in her ascending node at G, when the earth is at b; or, in her descending node, at E, when the earth is at a, what the effect would be.

PUPIL. She would be in a line with the sun.

TUTOR. And, on the sun's disc, she would appear a dark round spot, passing over it. These appearances, which are called transits, happen very seldom: because she is very seldom in or near her nodes at her inferior conjunctions. There was one in June 1761, one in June 1769; and the next will be in the year 1874. And as Mercury is seen in the same manner, it is a proof that their orbits must be within that of the earth.

PUPIL. I thank you, Sir, and shall be obliged to you to inform me how many constellations the earth pastes over in every revolution?

TUTOR. Twelve, which correspond with the months of the year, and are called the twelve signs of the zodiac.

PUPIL. What is the zodiac?

TUTOR. That part of the heavens which contains the twelve signs, and which you may conceive to be a zone or belt extending eight degrees on each side the ecliptic, in which the planets constantly revolve: so that no planet is ever seen more than eight degrees either north or south, that is above or below the ecliptic.

PUPIL. What am I to understand by a degree?

TUTOR. All circles, whether great or small, are supposed to be divided into 360 equal parts, called degrees, and each degree into 60 equal parts, called minutes: therefore, if I speak of a circle in the heavens, the circumference of the earth, or any other circle, by a degree is meant the 360th part of that circle; and a minute the 60th part of a degree.

PUPIL. What are the names of the twelve signs?

TUTOR. The first is called Aries, which you know signifies a Ram; Taurus, the Bull; Gemini, the Twins; Cancer, the Crab; Leo, the Lion; Virgo, the Virgin; Libra, the Balance; Scorpio, the Scorpion; Sagittarius, the Archer; Capricorn, the Goat; Aquarius, the Water-bearer; and Pisces, the Fishes.

PUPIL. Do you wish me to commit these to memory, Sir?

TUTOR. It is very requisite; but as I know you are fond of verse, you shall hear what Doctor Watts says—

The Ram, the Bull, the heav'nly Twins,

And next the Crab the Lion shines,

The Virgin, and the Scales:

The Scorpion, Archer, and Sea-goat,

The Man that holds the Water-pot,

And Fish with glitt'ring tails.

PUPIL. I like it much, as it will assist my memory.

TUTOR. As the twelve signs correspond with the months of the year, the earth must pass over nearly one degree every day, one sign every month, and in twelve months complete a whole circle, or 360 degrees; therefore every sign must contain 30 degrees, because 30 multiplied by 12 is equal to 360.

PUPIL. It must be so.

TUTOR. You must remember, that when the earth is in any sign, as seen from the sun, the sun will be in the opposite sign, as seen from the earth: for instance, if the earth be in Aries, the sun will be in Libra; if in Taurus, the sun will be in Scorpio, &c. therefore, as by the earth's annual motion, the sun *appears* to move, we always speak of the sun's, not the earth's place, in the ecliptic.—You do not seem to understand me?

PUPIL. Not perfectly, Sir.

TUTOR. Take this orange, and put it in the middle of the round table before us, and place an apple on the opposite side next the window: the orange may represent the sun, the apple the earth, and the window the sign Aries. Now go round the table to the apple; look at the orange, and tell me to what part of the room the eye will be directed.

PUPIL. To the part opposite to the window, Sir.

TUTOR. If then you suppose the door, which is opposite to the window, to be the sign Libra, the sun will be in Libra when the earth is in Aries—will it not?

PUPIL. It is very plain.

TUTOR. I shall now give you a table of the signs, their characters, the corresponding months, and the days of the month the sun enters each sign, by means of which, if you reckon a degree for a day, you may find the sun's place, nearly, for any day in the year.

PUPIL. This will give me much pleasure, and I shall be happy to have it.

THE TABLE.

NORTHERN SIGNS.

Aries,	Taurus,	Gemini,	Cancer,	Leo,	Virgo.
♈	♉	♊	♋	♌	♍
March,	April,	May,	June,	July,	Aug.
20,	20,	21,	21,	23,	23.

SOUTHERN SIGNS.

Libra,	Scorpio,	Sagittarius,	Capricorn,	Aqua.	Pisces.
♎	♏	♐	♑	♒	♓
Sept.	October,	November,	Decem.	Jan.	Feb.
23,	23,	21,	21,	20,	18.

PUPIL. Why do you write northern and southern signs, Sir?

TUTOR. Because they are situated north and south of a circle in the heavens, called the equinoctial, which circle crosses the ecliptic in the points Aries and Libra, and extends 23-1/2 degrees on each side of it; and which I shall have occasion to mention to you another time.

PUPIL. When you think proper, Sir, I shall be glad to have it explained to me.

TUTOR. Look at your table, and tell me what sign and what degree the sun is in the 30th of March, and 20th of October.

PUPIL. The sun enters Aries the 20th of March, of course he must be 10 degrees in that sign the 30th; and, as he does not enter Scorpio till the 23d of October, he must want three degrees of completing the sign Libra; he must therefore, on the 20th of October, be in 27 degrees of Libra.

TUTOR. Very well.—Do you learn the table, as you will have a farther use for it.

DIALOGUE VI.

Pupil.

Since I was last with you, Sir, I have been thinking of what you then told me, that the planets perform their revolutions in open space: I have not the least idea how this can be; if convenient, I shall be happy to have it explained.

TUTOR. It will be necessary first to inform you, that the orbits or paths described by the revolution of the planets round the sun, are not true circles (as Plate II. fig. 2.) but somewhat elliptical, that is, longer one way than the other, as fig. 3.

PUPIL. This is exceedingly plain.

TUTOR. In a circle, the periphery or circumference is equally distant from a point within called its center, as A; but an ellipsis has two points called the focuses or foci, as B C. In one of these, called its lower focus, is the sun: so that you see in every revolution of the planet it must be nearer to the sun in one part of its orbit, than it is in another.

PUPIL. I see it clearly.

TUTOR. Now let S (Plate II. fig. 4.) represent the sun, A B C D a planet in different parts of its orbit; when it is nearest to the sun, as at A it is said to be in its *perihelion*; when at B its *aphelion*; but when at C or D its middle or mean distance, because the distance S C or S D is the middle between A S the least and B S the greatest distance; and half the distance between the two focuses is called the *eccentricity* of its orbit, as S E or E F.

PUPIL. This I will endeavour to understand; but I find it will take me some time to be perfected in it.

TUTOR. You may study it at your leisure, as it will not prevent our proceeding to the thing proposed, namely, the laws which govern the motion of the planets, or ATTRACTION OF GRAVITATION.

PUPIL. By attraction I think you mean that property in bodies whereby they have a tendency to approach each other. I remember you told me that the magnet I had the other day attracted the needle.

TUTOR. Yes. And you may recollect that when I took a feather suspended by a thread, and put it near the conductor of the electrical machine, it was strongly attracted by it, and adhered to it as long as the machine was kept in motion.

PUPIL. I remember it well. But what am I to understand by attraction of gravitation?

TUTOR. The sun, being the largest body, *attracts* the earth and all the other planets, they *gravitate* or have a tendency to approach the sun; the earth being larger than the moon *attracts* her, and she *gravitates* towards the earth; the planets are attracted by and gravitate towards each other; a stone when thrown from the earth, by its attraction and the gravitating power or weight of the stone, is brought to the earth again; the waters in the ocean gravitate towards the center of the earth; and it is by this power we stand on all parts of the earth with our feet pointing to the center.

PUPIL. This information affords me great pleasure.

TUTOR. Having mentioned attraction of magnetism, electricity, and gravitation, it may not be amiss to inform you of another kind, called *attraction of cohesion*.

PUPIL. Any thing which tends to my improvement, I shall be obliged to you to communicate.

TUTOR. By attraction of cohesion is meant that property in bodies which connects or firmly unites the different particles of matter of which the body is composed.

PUPIL. Pray, Sir, inform me what you mean by the *laws* of attraction?

TUTOR. You are to understand, 1st. That *attraction decreases as the squares of the distances between the centers of the attracting bodies increase.*

PUPIL. I must beg you, Sir, to explain to me the meaning of the squares of the distances.

TUTOR. Any number multiplied into itself is a square number, thus 1 is the square of 1; 4 is the square of 2; 9 is the square of 3, and so on, because 1 multiplied into itself is 1; 2 by 2 is 4; 3 by 3 is 9, &c. Now suppose, that when the planet is at B (Plate II. fig. 4.) it is twice as far from the sun as it is at A: how much more will it be attracted by the sun at A than at B?

PUPIL. You say, Sir, that the distance is twice as great at B as at A?

TUTOR. I do.

PUPIL. Then as the square of the distance 2 is 4, the decrease of attraction at B, the planet at A will be attracted with four times the force it would be at B.—Am I right, Sir?

TUTOR. Perfectly so. And if the distance at B were three times as great as at A, it would be attracted with a force nine times as great.

PUPIL. I perceive it must be so.

TUTOR. I shall now give you the 2d law, namely, That *bodies attract one another with forces proportionable to the quantities of matter they contain.*

PUPIL. Do all bodies of the same magnitude contain equal quantities of matter?

TUTOR. No, certainly: For a ball of cork may be as large as one of lead, and yet not contain the same quantity of matter, because it is more porous, and not so compact or dense a body as the lead; neither will a ball of lead of the same magnitude as one of gold contain an equal quantity of matter.—So the sun, though a million of times as big as the earth, contains a quantity of matter only 200,000 as great, therefore attracts the earth with a force 200,000 as great as the earth attracts him.

PUPIL. I think this is clear.

TUTOR. We will now suppose that in the river are two boats of equal bulk, at the distance of twenty yards from each other, and that a man in one boat pulls a rope which is fastened to the other, what effect will be produced, or where do you think the boats will meet?

PUPIL. Had you not told me that bodies attract one another with forces which are proportioned to the quantities of matter they contain, I should say the boat to which the rope is fastened would come to that in which the man stands: but as I imagine you mean to apply this to attraction, by the above rule, they will meet at a point which is half way between them.

TUTOR. If one boat were three times the bulk of the other, how then?

PUPIL. The lightest would move three times as far as the heaviest, or 15 yards whilst the heaviest moved only 5.

TUTOR. Upon my word you reason philosophically. In both cases you are perfectly right.

PUPIL. As the sun is so immense a body that his quantity of matter is so much greater than the planets, I am at a loss to know why they are not by the power of attraction drawn to him.

TUTOR. And so they would if the attractive power were not counteracted by another of equal force.

PUPIL. Did you not say, Sir, that the planets are kept in their orbits by attraction?

TUTOR. I did. But you find that by attraction *only* the sun would draw all the planets to himself.

PUPIL. That is evident. But I wish to know what this counteracting power you speak of is?

TUTOR. I will tell you presently.—You must remember that *simple* motion is naturally rectilineal, that is, all bodies, if there were nothing to prevent them, would move in strait lines.

PUPIL. Then as the planetary motion is circular, it cannot be simple?

TUTOR. No. It is a *compound* of the two forces I have been mentioning: the one is called the attractive or centripetal force; the other, the projectile or centrifugal force.

PUPIL. The former I clearly comprehend, but not the latter. I can conceive, that if two bodies approach each other by attraction they must move in a right line.

TUTOR. If you shoot a marble on a smooth piece of ice, in what direction will it run?

PUPIL. Strait forward.

TUTOR. This is a projectile force.—Could you, do you think, shoot it in any other direction?

PUPIL. No, Sir.

TUTOR. Then is not this motion also rectilineal?

PUPIL. It is.

TUTOR. When you strike a ball with your cricket-bat, or throw a stone with your hand, is it not projected or thrown forward by the force of the bat or hand?

PUPIL. Certainly.

TUTOR. And does it not move in a strait line?

PUPIL. At first it appears to do so; but afterwards it inclines towards and falls to the earth.

TUTOR. Cannot you account for this?

PUPIL. I suppose it must be drawn to the earth by attraction.

TUTOR. You are right. The attraction of the earth, and the resistance of the atmosphere or air through which it moves, retards its progress, or it would continue moving in a strait line, with a velocity equal to that which was at first impressed upon it. In like manner the beneficent Creator of the Universe impressed a force on all the planets which should be equal to that of the attractive power of the sun, that one might not overcome the other.

PUPIL. This wants explaining.

TUTOR. I would willingly gratify you, but as I have much more to say on the subject, I fear it will be too great a burthen on your memory; it will therefore be better to postpone it.

PUPIL. As you please, Sir.

DIALOGUE VII.

TUTOR.

Having at our last meeting explained to you the nature of the attractive and projectile forces, I shall proceed to shew you that it is by the joint action or combination of these two forces that the planets are retained in their orbits.

PUPIL. I am all anxiety, as I wish to be informed how, or in what manner they can act against each other, to produce that effect.

TUTOR. Answer me a few questions, and you will soon know.

PUPIL. As many as you please, Sir.

TUTOR. If you whirl a stone in a sling, what will be its motion?

PUPIL. Circular.

TUTOR. Is you let it suddenly slip out of the sling, will it continue its circular motion?

PUPIL. No, Sir, but fly off in a strait line.

TUTOR. This line you must remember is what mathematicians call the tangent of a circle, as A a, B b, &c. (Plate II. fig. 5.) for all bodies moving in a circle have a natural tendency to fly off in that direction. Thus a body at A will tend towards a; at B towards b, and so on; but the central force acting against it preserves its circular motion.

PUPIL. By the central force here you mean the action of the hand, do you not?

TUTOR. Yes. For, as soon as the stone is released and that power is lost, it assumes its natural, that is, its rectilineal motion.—Again. If you are left at liberty, cannot you run strait forward?

PUPIL. Yes, Sir.

TUTOR. Now, suppose one of your companions were to fasten a rope round your body, and at the extent of it were to stand still and hold it tight, with a force equal to that with which you run, could you, do you think, move in a strait line, that is, in a tangent of a circle?

PUPIL. No, Sir. I must run in a circle.

TUTOR. Why?

PUPIL. Because, whilst the rope is extended I am prevented running in any other direction.

TUTOR. Just so it is with the planets: the attractive or centripetal force of the sun being equal to that of the projectile or centrifugal force of the planets, they are by attraction prevented moving on in a strait line, and, as it were, drawn towards the sun; and by the projectile force from being overcome by attraction. They must therefore revolve in circular orbits.

PUPIL. What I have so long wished is now accomplished. I understand it perfectly.

TUTOR. What I have now explained relates not only to the primary planets which have the sun for their center of motion; but, you must remember that the secondary planets are governed by the same laws, in revolving about their respective primaries; for, as by the attractive power of the sun combined with the projectile force of the primary planets they are retained in their orbits; so also the action of the primaries upon their respective secondaries together with their projectile force, will preserve them in their orbits.

PUPIL. Pray, Sir, what have you else to observe?

TUTOR. Have I not told you that the orbits of the planets are not true circles, but a little elliptical?

PUPIL. Yes, Sir; and I shall be glad to know the reason of it.

TUTOR. If the attractive power of the sun were uniformly the same in every part of their orbits they would be true circles, and the planets would pass over *equal* portions of their orbits in *equal* times; that is, they would move from B to C, (Plate II. fig. 5.) in the same time as from A to B, &c.

PUPIL. That is clear, but as their orbits are elliptical, when the planets are farthest from the sun, the velocity with which they move must be lessened as the attraction is decreased.

TUTOR. And they must consequently pass over *unequal* parts of their orbits in *equal* portions of time. And, as *a double velocity will balance a quadruple or fourfold power of gravity or attraction*, it follows, that as the centripetal force is four times as great at A as at B (Plate II. fig. 4.) the centrifugal force will be twice as great, and would carry a planet from A to *a* in the same time it would from B to *b*, and in its orbit from A to *c* as soon as from B to *d*, and thereby describe the area, or space contained between the letters A S *c*, in the same time as the area or space B S *d*. For according to the laws of the planetary motions, in their periodical revolutions, *they always describe equal areas in equal times.*

- 31 -

PUPIL. The orbits of the comets being very elliptical, the irregularity of their motions must be exceedingly great.

TUTOR. Great, indeed!—One of them passed so near the sun as to acquire a heat which Sir Isaac Newton computed to be two thousand times hotter than red hot iron.[12]

PUPIL. Astonishing! If they pass so near the sun, the centripetal force must act powerfully on the body of the comet.

TUTOR. And that force, you know, must be equalled by the projectile force; so you find they move when near the sun with amazing celerity.—But when arrived at their aphelion, where the influence of the sun is weak, what a transition!

PUPIL. Wonderful, indeed!—Their motion is excessively slow, and the sun must appear little more than a fixed star. Surely they cannot be inhabited, can they?

TUTOR. We cannot speak positively; but, as they differ so much from the planets, which we have reason to suppose are so, it is imagined they are designed for some purpose unknown to us.

PUPIL. When is the earth in its perihelion?

TUTOR. In December; and our summer half year is longer than the winter half, by about eight days.

PUPIL. I suppose this is occasioned by the inequality of the earth's annual motion.

TUTOR. It is; and this inequality is the cause of the difference of time between the sun and a well regulated clock; the latter keeps equal time, whilst the former is constantly varying.

PUPIL. I have often seen in the almanack clock fast, clock slow, but did not know the meaning of it: I imagine it is that the clock should be so much faster or slower than the time by the sun as is there mentioned.

TUTOR. It is: but there are tables calculated to shew the difference of time for every day in the year; so that if you know the exact times of the day by the sun, and have one of these tables, you will see what the time should be by the clock, to a second, which is not shewn in a common almanack.

PUPIL. In speaking of the annual or yearly motion of the earth, you have no where mentioned the cause of the seasons; will it be agreeable to do it now, Sir?

TUTOR. The vicissitudes of the seasons, the cause of day and night, &c. shall be the subject of future lessons: we shall find sufficient to employ us at present.

PUPIL. I think you told me just now that the earth is nearest the sun in December; that is our winter; this seems a little mysterious.

TUTOR. It may appear so to you now, by-and-by you will be of a different opinion. I shall explain this matter to you with that of the seasons, &c.

PUPIL. I fear I have interrupted you.—As you said you had sufficient employment for us, I shall be glad to know what it is.

TUTOR. Hitherto I have spoken of the sun's being fixed, and that the planets revolve about him as a center. Instead of which the sun and planets move round one common center, called the center of gravity.

PUPIL. What is this center of gravity?

TUTOR. Have you never seen a person raise a heavy weight by means of a long pole or leaver, which it was not in his power to lift without it?

PUPIL. Yes, Sir, and it excited my astonishment.

TUTOR. Now, suppose the weight to see raised to be 10 Cwt. and the prop on which the leaver rested 1 foot from the body to be raised; and the person at the other end of the leaver 10 feet from the prop; with what weight must he press to raise the 10 Cwt.?

PUPIL. I think that very easy; for, as he is ten times as far from the prop as the weight is, a pressure of 1 Cwt. which is one-tenth of the weight to be raised will do it.

TUTOR. To be sure; and yet you say you were astonished when you saw it! Every thing we do not understand at first appears difficult.—To apply this to our present purpose. You see that a weight of 1 Cwt. at 10 feet from a prop, will balance another of 10 Cwt. at one foot from it. Now, instead of a prop let the two weights be nicely poised on a center, round which they may freely turn; the heaviest would move in a circle, whose radius, or distance from the center would be one foot, whilst the lightest would move in one 10 feet from the center in the same time.

PUPIL. Is the center round which they move the center of gravity?

TUTOR. It is; and round an imaginary point as a center the sun and planets move, always preserving an equilibrium. If the earth were the only attendant on the sun, as his quantity of matter is 200,000 times as great as that of the earth, he would revolve in a circle a 200,000th part of the earth's distance from him, in the same time as the earth is making one revolution in its orbit, or in one year; but, as the planets in their orbits must vary in their positions, the center of gravity cannot be always at the same distance from the sun.

PUPIL. If it were, the balance could not be preserved.

TUTOR. Clearly so. But you must know that the quantity of matter in the sun so far exceeds that of all the planets together, that even if they were all in a line on one side of him he would never be more than his own diameter distant from his center of gravity; therefore, astronomers consider the sun as the center of the system, and express themselves accordingly.

PUPIL. As you told me the secondary planets are governed by the same laws as the primaries, I imagine they also with their primaries move round a center of gravity.

TUTOR. They do so.—The earth and moon, Jupiter with his satellites, Saturn and his attendants, revolve about their respective centers; these, with the sun and the rest of the planetary system, make their circuits round their center; every system in the universe is supposed to revolve in like manner; and all these together to move round one *common center*.—How are we lost in contemplating the omniscience of the Deity! How difficult to conceive so many millions of bodies of dead matter constantly in motion, so nicely balanced and governed by such unerring laws!—Well may we say with the Psalmist, "Lord! how manifold are thy works, in wisdom hast thou made them all."

---

12. Dr. Herschel is of opinion, that bodies near the sun do not acquire so great a degree of heat as has been generally imagined.

DIALOGUE VIII.

Tutor.

I shall now, agreeably to my promise, explain to you the cause of day and night, and then proceed with the vicissitudes of the seasons.

PUPIL. That is what I much wish to know; and had you not told me that the earth moved round the sun every year, I should have found no difficulty in accounting for the succession of day and night, since the sun appears to rise and set every day.

TUTOR. That is true; but I think I must have convinced you that so immense a body as the sun cannot revolve about the earth; as well may you suppose that in roasting a bird it is necessary that the fire should move round it.

PUPIL. That I think would be very absurd, as it is much easier for the bird on the spit to turn to the fire, than for the fire to go round the bird.

TUTOR. You are certainly right, and if the earth revolve on its axis every twenty-four hours, will not the different parts of it be alternately turned to the sun, as the bird on the spit is to the fire?

PUPIL. I do not clearly comprehend what you mean by the axis of the earth; for, as it moves in open space and has no support, it can have nothing to resemble the spit on which it turns.

TUTOR. Certainly not. By the earth's axis is meant an imaginary line passing through its center, on which it is supposed to turn; as your ball if rolled on the ground would revolve on an axis whilst it was moving forward.

PUPIL. I can now answer your question in the affirmative: and, as our year consists of 365 days, I imagine the earth must make as many revolutions on its axis whilst it is going once round the sun.

TUTOR. Undoubtedly: and as only one half of a spherical body can at any time be enlightened by a luminous body, that part of the earth only which is turned to the sun, can receive the benefit of his enlivening rays, when it will be day; whilst the opposite part will be involved in darkness, and it will be night.

PUPIL. I perceive it must be so. But, if the earth move in the manner you describe, I cannot conceive how it is that we are not sensible of its motion.

TUTOR. If the motion of the earth were irregular it would be perceptible; but as it meets with no obstruction the motion must be so uniform as not to be perceived.

PUPIL. Had I recollected this, I need not have given you this trouble.—But I am continually meeting with fresh difficulties.

TUTOR. You have only to mention what they are, and I shall take a pleasure in removing them.

PUPIL. I thank you, Sir; and shall be obliged to you to inform me, how the motion of the earth can cause the sun to appear to move?

TUTOR. When in a carriage which went smoothly on the road, or in a boat whose motion was scarcely perceptible on the water, did you never fix your attention on the objects you passed?

PUPIL. Yes, often, Sir.

TUTOR. And had you not known that you really moved, and that the trees, &c. were immoveable in the ground, what then would have been your opinion?

PUPIL. That the trees, &c. moved in a direction contrary to that in which I was moving.

TUTOR. Is not this sufficient to convince you that the apparent motion of the sun may be occasioned by the revolution of the earth on its axis?

PUPIL. It is:—But if so large a body as the earth make a revolution on its axis in 24 hours, it must move with great velocity.

TUTOR. It does so; and the inhabitants of London by this motion are carried at the rate of 560 miles an hour[13].

PUPIL. What an astonishing rapidity!

TUTOR. Now, the sun with the rest of the heavenly bodies must move round the earth, or the earth must revolve on its axis in 24 hours, to cause that appearance.

PUPIL. That is plain.

TUTOR. Well then, great as you may suppose the velocity of the earth on its axis to be, if the sun move round the earth his hourly motion will be nearly 25 millions of miles; and beyond conception would be that of the fixed stars. Which now do you think is most probable, that the sun and stars should move round the earth, or that they, by the simple motion of the earth, should appear to be in motion?

PUPIL. The latter, to be sure, Sir.—I have one difficulty remaining, which is this; if a lark rise from a field near London and remain in the air a quarter of an hour, if the earth move at the rate of 560 miles an hour, it will go 140 miles whilst the lark is suspended, and yet it continues over the field,—how can this be?

TUTOR. This objection to the motion of the earth has been made by those who were older and who thought themselves wiser too than yourself. They either did not know or did not consider, that the atmosphere which surrounds the earth is a part of itself, and gravitates towards it, and therefore partakes of the earth's motion and carries the lark along with it. Besides, as the Sun, Venus, Mars, and Jupiter are known to revolve on their axes, we have reason to suppose that the other planets, together with the earth, must have the same motion[14].

PUPIL. How is it known that they do revolve on their axes; and in what time do they perform their revolutions?

TUTOR. By the assistance of telescopes dark spots have been seen on the disc of the sun, by the motion of which it is found that he revolves on his axis in 25-1/4 days; Venus performs her diurnal revolution in about ho. min. / 23.21; Mars goes round his axis in ho. min. / 24.39; and Jupiter in ho. min. / 9.56; as to the rest, no spot or any fixed point has been discovered to ascertain the length of their day; Mercury being too near the sun, and Saturn and the Georgium Sidus too remote for our observations.

PUPIL. I can no longer doubt of the earth's motion: and, if it will not be improper, a description of the atmosphere will give me pleasure.

TUTOR. That I can have no objection to. The atmosphere is a thin, invisible fluid, most dense or heavy near the earth, but grows gradually rarer or lighter the higher we ascend, so much so, that at the tops of some high mountains it is difficult to breathe. It serves not only to suspend the clouds, furnish us with wind and rain, and answer the common purposes of breathing, but is also the cause of the morning and evening twilight, and of all the glory and brightness of the firmament.

PUPIL. How, pray?

TUTOR. If there were no atmosphere, the sun would yield no light but when our eyes were directed towards him; and the heavens would appear dark and as full of stars as on a dark winter's night; but the atmosphere being strongly illuminated by the sun, reflects the light back upon us, and makes the whole heavens to shine so strongly, that the faint light of the stars is obscured, and they are rendered invisible.

PUPIL. I find then the atmosphere is of more use than I imagined. But how is it the cause of the twilight?

TUTOR. The atmosphere is about 45 miles above the surface of the earth, therefore the sun's rays falling upon the higher parts of it before rising, by reflection causes a faint light, which increases till he appears above the horizon; and in the evening it decreases after he sets, till he is 18 degrees below the horizon, where the morning twilight begins, and the evening twilight ends.

PUPIL. By the horizon, I think you mean that distant boundary of our sight where the heavens and the earth seem to join all around us, as it appears from an eminence.

TUTOR. The very same. 'Tis that imaginary circle which intercepts from our view the sun, moon, and stars each night; and when, by the rotation of the earth, they appear to descend below it, we say they are set; as on the contrary, each morning, when they appear above it, we say they rise.

"To find the spacious line, cast round thine eyes,

"And where the earth's high surface joins the skies,

"Where stars first set, and first begin to shine,

"There draw the fancy'd image of this line."

PUPIL. A very pleasing description, indeed.

TUTOR. You will remember that this is called the *rational horizon*; but that which respects land and water is called the *sensible horizon*. The former divides the heavens into two equal parts, and is 90 degrees distant from a point directly over our heads, called the *zenith*, and the opposite point of the heavens directly under our feet, called the *nadir*.—But I must resume the subject of the atmosphere.

PUPIL. Had I not thought you had finished your description of the atmosphere, I should not have presumed to interrupt you.

TUTOR. What I have told you respecting the horizon is necessary for you to be acquainted with; therefore, the suspension is immaterial.—You must, I make no doubt, have observed the sun and moon at rising and setting to appear larger than when higher above the horizon.

PUPIL. I have, frequently, Sir.

TUTOR. And cannot you tell the reason of it?

PUPIL. No, Sir.

TUTOR. The reason is this: In viewing them, when near the horizon, you see them through a thicker medium than when they are higher, that is, you see them through a greater quantity of the atmosphere; and you not only see them larger, but really above the horizon whilst they are actually below it.

PUPIL. How do you account for this, Sir?

TUTOR. Light, like material bodies, if it meet with no obstruction, will move in right lines; now, the rays of the sun in coming to the earth must pass through a great quantity of the atmosphere, which being a fluid, refracts or bends the rays of light, by which refraction it is that we are favoured with the sight of the sun 3-1/4 minutes every morning before he rises above the horizon, and every evening after he sinks below it, which in one year amounts to more than 40 hours. This refraction is greatest near the horizon, and ends in the zenith.

PUPIL. Pray, Sir, can you make this clearer by an experiment?

TUTOR. I have just thought of one. Take a bason filled with water, and a strait stick or piece of wire; put it perpendicularly into the water, that is, that it lean neither way, and there will be no refraction; incline it a little towards the edge of the bason and it will appear a little bent at the surface of the water; incline it still more, and the refraction will be greater.

PUPIL. I have often seen this appearance when I have put my stick into water, but did not before know the cause.

TUTOR. You may try one more experiment. Pour the water out of the bason, and set the bason on the floor; put a guinea into it, and let it represent the sun.—Why do you smile?

PUPIL. Because I have not the sun's representative to try the experiment with.

TUTOR. Well, well, put a shilling into the bason and call it the moon, and it will answer the same purpose:—Walk backward till you just lose sight of it, then the right line from your eye continued over the edge of the bason must pass beyond the money at the bottom of it.

PUPIL. That is evident.

TUTOR. Keep your position, and desire some friend to pour the water gently into the bason so as not to remove the money, and you will clearly distinguish it. Now, if you call the edge of the bason the horizon, the water the atmosphere, and the shilling the moon, is it not clear that you will see it above the horizon, when it is really below it?

PUPIL. I think so, Sir.

TUTOR. Well, try the experiment, and let me know the result when I next see you.

13. The hourly motion under the equator is 900 miles.

14. Dr. Herschell says that several of the fixed stars revolve on their axes.

DIALOGUE IX.

TUTOR.

I presume, Sir, you have made the experiment I recommended to you.

PUPIL. I have, Sir; and am so well convinced of what you told me, that nothing farther need be said on the subject.

TUTOR. As that is the case, I shall proceed.—I dare say you do not forget what the plane of the ecliptic is.

PUPIL. I do not, Sir; but have a perfect recollection of it.

TUTOR. Now, remember, that the axis of the earth is not upright or perpendicular to the plane of the ecliptic, but inclines to, or leans towards it, 23-1/2 degrees, and makes an angle with it of 66-1/2 degrees.

PUPIL. An angle signifies a corner; but that cannot be the meaning here.

TUTOR. That is what is generally understood by an angle: but, in geometry, it means the meeting of any two lines which incline to one another, in a certain point. Now, if you conceive the axis of the earth to be one line, and the plane of the ecliptic the other, the point where they meet or cross each other will form an angle.

PUPIL. I think I understand it; but how can it contain 23-1/2 or 66-1/2 degrees?

TUTOR. You know what a degree is.

PUPIL. If I remember right it is the 360th part of a circle.

TUTOR. It is so: and the measure of an angle is an arc or part of the circumference of a circle, whose angular point is the center: and so many 360th parts as any arc contains, so many degrees the measure of the angle is said to be; thus, Z C P (Plate III. fig. 1.) makes an angle of 23-1/2 degrees, because the arc Z P contains 23-1/2 360th parts of the whole circle. Then if A B represent the plane of the ecliptic, and N C S the axis of the earth, as D N contains the same number of degrees as Z P, will not its inclination from a perpendicular be 23-1/2 degrees?

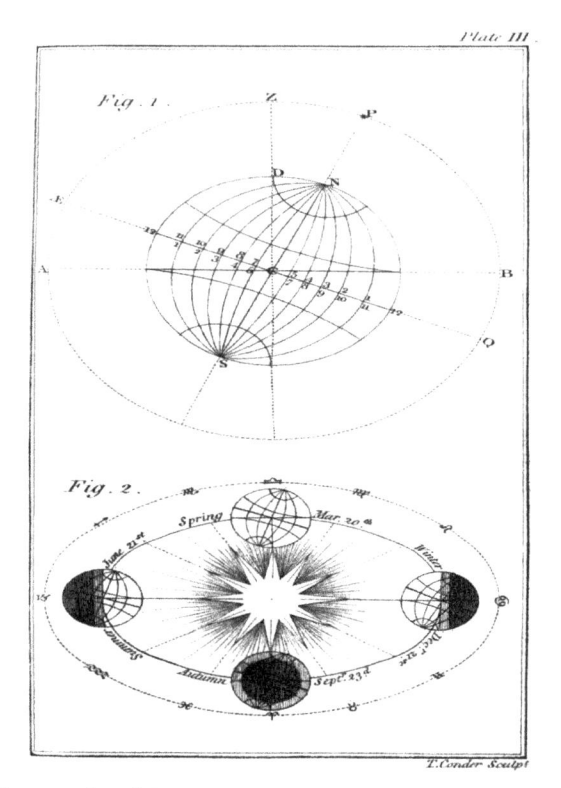

PUPIL. Nothing can be plainer.

TUTOR. For the same reason, as P B contains 66-1/2 parts of the whole circle, the axis of the earth makes an angle of 66-1/2 degrees with the plane of the ecliptic. And, if you add 23-1/2 to 66-1/2 the sum will be 90, which is the measure Z B, or the fourth part of the circle, and makes what is called a right angle, at the point or center C.

PUPIL. It is very clear:—but what do the other letters refer to?

TUTOR. The extremities of the earth's axis are called the poles, N the north, and S the south pole, and P the north-pole star, to which, and to the opposite part of the heavens, the axis always points. These extremities in the heavens appear motionless, whilst all other parts seem in a continual state of revolution: the circle of motion appears to increase with the distance from the apparently motionless points to that circle in the heavens which is at an equal distance between them, called the equinoctial, represented by the letters Æ Q; and is the same I promised some time ago to explain to you.

PUPIL. I recollect it: and as the line A B represents the plane of the ecliptic, I suppose the line Æ Q is the plane of the equinoctial, which I see crosses it as you then told me.

TUTOR. You are right: and it makes an angle with it of 23-1/2 degrees. It is called the equinoctial, because when the sun appears there, that is, in Aries or Libra, the days and nights are equal in all parts of the world, which I shall shew you in due time; and shall now explain to you what I have just mentioned, that the axis of the earth always points to the same parts of the heavens. I am apprehensive you will think it strange that this should be the case, and the axis keep parallel to itself.

PUPIL. What am I to understand by the axis being parallel to itself?

TUTOR. Two lines are said to be parallel when they do not incline to but keep at equal distances from each other; so that if they were infinitely continued, they would never meet. Now, if you can conceive a line drawn parallel to the earth's axis in any part of its orbit, it will be parallel to it in every other part of it. A little drawing I have by me, (Plate III. fig 2.) where the earth is represented in four different parts of its orbit, I think will make this plain to you.

PUPIL. I comprehend your meaning clearly. But, as the orbit of the earth is 190 millions of miles in diameter, I have not the least conception how it can incline to the same points. Had you not told me to the contrary, I should have thought it must move round them in every revolution of the earth about the sun.

TUTOR. That such a motion would be perceptible is evident, if the fixed stars were near the earth; but, compared with their distance, 190 millions of miles is but a mere point: therefore, the axis always inclines to the same points of the heavens.

PUPIL. This is a greater proof of the inconceivable distance of the stars than what you mentioned before, and I thought that very astonishing:

Wonders on wonders constantly arise,

Whene'er we view the earth, or sea, or skies.

TUTOR. It is very true. And the more we search, the more we have cause to admire the works of the Almighty.

PUPIL. Pray, Sir, what is the next thing you propose?

TUTOR. To make you acquainted with the other circles you see in the figure (Plate III. fig. 1.) as it is very necessary you should know them.

PUPIL. Will you be kind enough to tell me their names, Sir, and I will endeavour to remember them?

TUTOR. That line which divides the globe into two equal parts, called the northern and southern hemispheres, which answers to the equinoctial in the heavens, and is equally distant from the two poles, is called the *equator*; the other which crosses it, as I before told you, is the *ecliptic*; the smaller circle, north of the equator, is the *tropic of Cancer*; that south of it, the *tropic of Capricorn*; the circles next the poles are called the *polar circles*; or that next the north pole, the *arctic circle*, and that next the south pole, the *antarctic circle*; each of which is 23-1/2 degrees distant from its respective pole, as are the tropics from the equator.

PUPIL. You have not mentioned the lines which cross the other circles, and terminate in the poles; what are they called?

TUTOR. They are called *meridians*, because when any of them, as the earth revolves on its axis, is opposite to the sun, it is mid-day or noon along that line. Twenty-four of these lines are usually drawn on the globe to correspond with the twenty-four hours of the day; but you are not to suppose there are no more than twenty-four; for every place that lies ever so little east or west of another place has a different meridian.—To make this clearer to you, we will suppose the upper 12 (Plate III. fig. 1.) to be opposite the sun, it will of course be noon along that line; the next meridian marked 1, being 15 degrees east, will have passed the meridian 1 hour, consequently it will there be one in the afternoon, and so on, according to the order of the figures, till you come to the lower 12, which being the part of the earth turned directly from the sun, it will be midnight on that meridian; on the next meridian, as you proceed round, it will be one in the morning, the next two, and so on till you arrive at the upper twelve, where you set off. So you see there must be a continual succession of day and night. This difference of time between places lying under different meridians is what is called longitude.

PUPIL. I think I have heard of a Mr. Harrison, who made a time-keeper for determining the longitude. Shall I trespass at all if I beg a little farther information on this subject?

TUTOR. It is my wish at all times to satisfy your curiosity, when I can do it with propriety. I shall therefore comply with your request.—Mr. Harrison's time-keeper, and those made since by other artists, are so constructed, that the heat and cold of different climates will not affect them; for, all metals are more or less expanded by heat, and contracted by cold; for which reason it is, that a clock or watch made in the usual way will not keep equal time. Now, all that is required of these time-keepers to ascertain the longitude is this: Suppose a captain of a vessel sailing from London to the

West Indies, we will say Kingston, in Jamaica. On his passage thither he makes an observation, and finds the sun on the meridian, or that it is twelve o'clock in that situation, when by his time-keeper it is two in the afternoon in London, whence he concludes he is 30 degrees west of London.

PUPIL. I must beg you to explain this to me, as I do not understand why two hours of time should be equal to 30 degrees of longitude.

TUTOR. You must consider, that as the earth makes a complete revolution on its axis in 24 hours, it must pass over 360 degrees in that time: now, if you divide 360 by 24, the quotient 15, will be the number of degrees passed over in one hour; 30 degrees will be equal to two hours, &c. The difference of time between London and his situation is two hours, consequently the difference of longitude must be 30 degrees: and, it must be west, because the sun had passed the meridian of London; for, as the earth revolves from west by south to east, one place which lies east of another must come first to the meridian or opposite to the sun. Therefore, when longitude is reckoned from London, if the place lie east of that meridian the time will be before; if west, after London.

PUPIL. I see it clearly; and as 60 minutes make an hour, if I divide it by 15, the quotient 4 will be the minutes answering to one degree.

TUTOR. You are right: and for the same reason, 4 seconds of time are equal to one minute of longitude, which you know is the 60th part of a degree.— Our captain when arrived at Kingston, finds the difference of time between it and London 5 ho. 6 min. 32 sec. Can you tell me the longitude of Kingston?

PUPIL. If I bring the hours and minutes to minutes, and divide by 4, the quotient I think will be degrees, will it not?

TUTOR. It will: and the seconds of time divided by 4, will be minutes of longitude. Now try if you can do it.

PUPIL. Five hours 6 minutes, multiplied by 60 will be 306 minutes, this divided by 4, will give 76 degrees and 2 over, which 2 is half a degree, or 30 minutes: and 32 seconds of time divided by 4, will be 8 minutes of longitude, the sum of which is 76 degrees 38 minutes for the longitude of Kingston.

TUTOR. Very well.—I have just now thought of another method of reducing time to longitude, and longitude to time, which you may probably find easier. However, when you are in possession of both, you may use which you please.

PUPIL. That which is easiest must, I think, be best.

TUTOR. I will give it you, and let me have your opinion of it.

To reduce time to longitude.

Multiply the hours, minutes, and seconds of time by 15, or rather by the factors as they are called, namely 3 and 5, carrying one for every 60 in the minutes and seconds, and setting down the remainder, thus:

	ho.	min.	sec.	
	5	6	3	difference
			2	of
			3	time.
	1	1	3	
	5	9	6	
			5	
Degrees	7	3	0	longitude.
	6	8		

Divide the degrees and minutes of longitude by 5 and 3 and the quotient will be the difference of time.

PUPIL. I give this the preference.

TUTOR. As longitude is seldom mentioned without being accompanied with latitude, that you may not be ignorant of its meaning when you meet with it, I shall just tell you that it is the distance of any place from the equator, reckoned in degrees and minutes on the meridian, and is either north or south as the place lies north or south of the equator. The latitude of any place is equal to the elevation of the pole above the horizon. The latitude of the heavenly bodies is reckoned from the ecliptic, and terminates in the arctic and antarctic circles: and their longitude begins at the point Aries.

PUPIL. What is the measure of a degree?

TUTOR. A degree of latitude is 60 geographical, or 69-1/2 English miles: and a degree of longitude on the equator is equal to it, because the equator as well as the meridians divides the globe into two equal parts. But a degree of longitude decreases as you approach the poles: for at the poles the meridians meet in a point, consequently a degree there can have no dimension. To-morrow I will shew you the cause of the seasons.

DIALOGUE X.

PUPIL.

I think, Sir, when you left me last night you told me our next business would be to explain the nature of the seasons?

TUTOR. I did so, and am persuaded you will find no great difficulty in comprehending it.—Cast your eye on the little drawing I gave you, (Plate III. fig. 2.) where the earth is represented as situated at the four quarters of the year, namely, Spring, Summer, Autumn, and Winter.—But before we proceed to an explanation it will be necessary to remark, that, in the little scheme the eye is supposed to be elevated above the plane of the earth's orbit, and that we see it very obliquely. The orbit by this means appears very elliptical; and, the enlightened hemisphere, or that half of the earth which is turned to the sun in the spring, and the darkened hemisphere, or that turned from him in the autumn, are there represented.

PUPIL. This I understand.

TUTOR. Well then, we will begin with the spring.—In this situation of the earth the equator is exactly opposed to the sun: and, as he always enlightens a hemisphere, or half of its surface, his rays will reach to both the poles: whence, from the diurnal revolution of the earth, the day and night are equal all over the globe.

PUPIL. This I remember you told me happened when the sun was in Aries and Libra. The sun is now entering Aries: and, as we are in the rays of the sun one half of the diurnal revolution, and in the shadow of the earth, or dark, the other half, the day and night must be equal.

TUTOR. Certainly. And as the sun enters Aries in the equinoctial, it is then called the *Vernal*, that is, *Spring Equinox*. When the sun enters the opposite sign Libra, the same effects are produced, and it is then called the *Autumnal Equinox*.

PUPIL. You have passed on from Spring to Autumn.

TUTOR. I have so.—We will now return, and trace the earth in its orbit from spring to summer.—You have already seen that the north and south poles are both enlightened, and that the day and night are equal at the equinoxes. If the axis of the earth were perpendicular to the plane of the earth's orbit, this would constantly be the case, and we should have no diversity of seasons: for, the sun being over the equator, the poles must be perpetually enlightened, and of course we should have equal day and night at all times of the year.

PUPIL. That is plain. I suppose then that it is to the inclination of the earth's axis we are indebted for the increase and decrease of days.

TUTOR. It is occasioned by the inclination of the earth's axis and its preserving its parallelism, which I explained to you last evening.—As the sun is now in the first point of Aries, the earth you know must be in the beginning of Libra, it being the opposite sign.—Now fix your attention on the scheme, and imagine the earth to be advancing in its orbit through Libra, Scorpio, and Sagittarius: and at the first degree of Capricorn give me your opinion of the earth's position.

PUPIL. The north pole is turned to the sun, the south pole from him, and the tropic of Cancer is opposite to him.

TUTOR. How many degrees are the tropics from the equator, or, in other words, what is the inclination of the earth's axis?

PUPIL. Twenty-three degrees and a half.

TUTOR. And so far are the rays of the sun cast beyond the north pole, and fall short of the south pole: so that the whole of the arctic circle is enlightened, and the antarctic circle involved in darkness.

PUPIL. What conclusion am I to draw from this?

TUTOR. That in the northern half of the globe it is the longest day, or summer, and in the southern half the shortest, or winter, whilst under the equator the days and nights are equal.

PUPIL. I used to think that when it was winter or summer here it was so in every part of the world.

TUTOR. You now find your mistake. For as the earth is making its progress from Libra, the north pole is approaching the sun, and the south pole receding from him: consequently the length of the day is increasing in the northern hemisphere and decreasing in the southern.—The sun has now been three months above the horizon of the north pole, and the same time below that of the south pole, and in three months more, when the earth arrives at Aries, the scene will be reversed: the sun will be over the equator, both poles will be again enlightened, and the day and night will be equal in every part of the globe. The sun will now be rising to the south and setting to the north pole. This is our Autumn.

PUPIL. And as the earth is advancing towards winter, the south pole will be turning to the sun, and the north pole from him, whence I conclude that when the earth is in Cancer it must be summer, south of the equator, when it is our winter.

TUTOR. Most assuredly. For you see that the sun is over the tropic of Capricorn, which you know is as much south of the equator as the tropic of Cancer is north of it, where the sun was in our summer. The antarctic circle is now enlightened, and the arctic obscured in shade; but, under the equator there is neither increase nor decrease, the days and nights being each twelve hours.

PUPIL. It is now our winter, the sun has been three months above the horizon of the south pole, and will continue so till the vernal equinox, when he will again rise to the north pole, and so on in regular succession.

TUTOR. It must be plain then to you that there can be but one day and one night at each of the poles, reckoning the time the sun is above or below their respective horizons; under the arctic and antarctic circles, the longest day is twenty-four hours, and in the shortest the sun is just visible in the horizon at noon. The longest day decreases in length the nearer we approach the equator, where I before observed there is no variation, because the circle bounding light and darkness, in every position of the earth, divides the equator into two equal parts; and, it must be observed, that the longest day and longest night are equal to each other in every part of the globe.

PUPIL. If the longest day under the arctic circle be just twenty-four hours, the sun must rise in the north.

TUTOR. He does so, makes a complete circle and sets in the [15]north again. From the arctic circle to the equator, he rises north of the east and sets north of the west: at the equator he rises due east and sets due west, thence southward to the antarctic circle, he rises south of the east, and sets south of the west: and under the antarctic circle, as I observed just now, he is visible in the horizon in the south at noon.

PUPIL. We usually say, the sun rises in the east and sets in the west.

TUTOR. At the equinoxes it must be so in all parts of the globe, the poles excepted: in every other situation, except under the equator, there is a continual change. What I have now told you, respecting the northern hemisphere, will be reversed at our shortest day: that is, in the northern hemisphere the sun will rise south of the east and set south of the west; and, in the southern hemisphere the contrary, the sun will be in the horizon, at noon, under the arctic circle, and the day will be twenty-four hours under the antarctic circle.

PUPIL. Pray Sir, are the regions within the polar circles inhabited? If they are, their situation, in winter, must, I think, be dreadful.

TUTOR. It is foreign to my present purpose to speak of the inhabitants of the earth, as that more properly belongs to Geography. Thus much however I shall tell you, that, although it must be very cold and dreary, they are not so long deprived of light as you may imagine; for, even under the poles, when the sun is hidden from them, they are but a short time in total darkness, for, you must recollect, that the twilight continues till the sun is eighteen degrees below the horizon; and the sun's greatest depression, you know, can be but twenty three degrees and a half, equal to the inclination of the earth's axis. Besides this, the moon is above the horizon of the poles a fortnight together; being half her period north, and the other half south, of the equator; and, as the moon at full is in the sign opposite to the sun, the tropical full moons must be twenty-four hours above the horizon at the polar circles.

PUPIL. This description is very pleasing, as I had no idea of their being favoured with so much light in the absence of the sun: and, I find, as the sun is longer above the horizon in summer than in winter, the moon, on the contrary, continues longer with us in winter, when we most need her assistance, than she does in summer.

TUTOR. As you seem to understand what I have been explaining, I shall shew you, that the reason why it is hottest when we are farthest from the sun is, that in winter when we are nearest to him the days are shorter, his rays sail very obliquely on us, and are more dispersed than they are in summer, when he not only remains longer above the horizon, but being higher, his rays fall more direct on us, by which means the earth becomes so much heated that it has not time in the short nights to get cold again.— When the earth is nearest the sun it is summer in the southern hemisphere, therefore it is reasonable to suppose that the heat there must far exceed ours in the same latitude; but to counteract this their summer is shorter by eight days than ours: and it is well known that it is much colder near the poles in the southern than in the northern hemisphere: but this is accounted for from there being more land to retain the heat in the latter than in the former.

PUPIL. My doubts on this head being now removed, I must beg you to give me such other information as you may think proper.

TUTOR. As there are different degrees of heat and cold, the earth has been divided into five zones, namely, one torrid, two temperate, and two frigid zones.

PUPIL. How are they distinguished?

TUTOR. The torrid zone is all that space surrounding the globe contained between the tropics, having the equator running through the middle of it. It is so called on account of its excessive heat, for, twice every year the sun is vertical to the inhabitants, that is, he shines directly on their heads, and casts no shadow, but under their feet, at noon.

PUPIL. We find it sometimes extremely hot here in our summer; surely, in the torrid zone it must be almost insupportable?

TUTOR. They are inured to it from their infancy.—But we are departing from our subject.—The temperate zones are comprehended between the tropics and polar circles, that between the tropic of Cancer and the arctic circle is called the north temperate zone, and that between the tropic of Capricorn and the antarctic circle the south temperate zone.

PUPIL. I suppose they are called temperate because the heat is not so intense as in the torrid zone?

TUTOR. True. Neither is the cold so severe as in the frigid zones, which are those regions comprized within the polar circles, and are denominated north and south, as they are contiguous to the north or south poles.

PUPIL. Why are they called frigid?

TUTOR. They are called frigid or frozen zones, because near the poles there are perpetual fields of ice, the heat of the sun, even in summer, being insufficient to dissolve it.—Now try if you can tell me the breadth of each zone in degrees.

PUPIL. The torrid zone being twenty-three degrees and a half on each side the equator must be forty-seven degrees, which must also be the breadth of the frigid zones, as the polar circles are distant twenty-three degrees and a half from the poles, which are their centers. And, as from the equator to either pole is ninety degrees, from the equator to the tropics twenty-three and a half, and from the polar circle to the pole twenty-three and a half, if the sum of these, that is, forty-seven, be taken from ninety, the remainder, forty-three, will be the breadth of each of the temperate zones.

TUTOR. Very well.

PUPIL. From what you have told me I have no doubt but that the earth is globular, but I have no proof of it: I must therefore beg your assistance.

TUTOR. That it cannot be an extended plane, as some have imagined, is very evident; for, if it were, the angle made with that plane and the north pole star would be always equal, for reasons I have before given you: neither can it be cylindrical, that is like a garden roller, as others have supposed.—If a person travel northward the pole star becomes more elevated, and if he could penetrate to the north pole of the earth the star would be in the zenith, or directly over his head: on the contrary, if he travel southward, it is more and more depressed till he arrives at the equator, where the star is in the horizon; as he proceeds it disappears, and other stars rise to his view, invisible to us. Here then you see it must be circular northward and southward.

PUPIL. I am convinced it must be so.

TUTOR. And it is as certain that it is so east and west: for, navigators have often sailed round it steering the same course: that is, if they sail an easterly or westerly course at setting off, by continuing the same course they will return to the port whence they departed. This you know they could not do if it were not round, any more than an insect could, by crossing a round table, arrive at the place it set out from; but, by going round the edge it would be still going forward and come again to the point it had left.

PUPIL. It is very evident.

TUTOR. Again. In every direction, if a ship be seen at a distance, the first things observed are the top-mast and rigging, whilst the hull or body of the ship is hidden behind the convexity, that is roundness of the water, just as you would see a man coming over a hill, you would first see his head, he would be rising more and more to your view till he arrived at the top, where he would be full in sight.

PUPIL. I am at a loss to account for the convexity of the water. How can its surface be round?

TUTOR. Have you never observed the drops of water falling from the eaves of a house?

PUPIL. Often, Sir.

TUTOR. Of what shape were they?

PUPIL. Globular.—But what is the cause of their being so?

TUTOR. Attraction.—For as every particle of water which composes the drop tends to the same center, every part of the surface must be equidistant from the center, it must therefore be spherical. In like manner if you separate quicksilver, each portion will form itself into a globe.

PUPIL. All this is very clear. And, for the same reason, the water in the ocean must be convex; for, I remember you told me that it gravitated towards the center of the earth.

TUTOR. Once more.—I think you must have seen an eclipse of the moon.

PUPIL. I have, Sir.

TUTOR. Of what figure was the darkened part?

PUPIL. Circular.

TUTOR. Take this ball, and hold it before the candle between your finger and thumb, so that the shadow may be thrown on the wall, and in all positions you will find it circular.

PUPIL. It is so.

TUTOR. Apply this crown piece in the same manner, with the flat side to the candle.

PUPIL. It is a circle.

TUTOR. Turn it a little obliquely.

PUPIL. It is now an ellipsis.

TUTOR. Now turn the edge to the candle.

PUPIL. The shadow is a strait line.

TUTOR. You now see that no other body than that of a globe can in all positions cast a circular shadow.

PUPIL. I do, Sir.

TUTOR. The darkness on the disc of the moon at the time of an eclipse is the shadow of the earth, which in all situations is circular; the earth, therefore, which casts the shadow, must be a globe.

PUPIL. It must be so.—But——

TUTOR. The earth is mountainous.—It is so: but remember that the highest mountain bears no greater proportion to the bulk of the earth than the small irregularities on the peel of an orange bears to that fruit: that objection therefore is soon removed. And yet it is not a true sphere.

PUPIL. What then?

TUTOR. A spheroid, that is, it is a little flattened at the poles, and is in shape not unlike an orange or a turnip. This you will not be surprised at when I tell you that the equatorial parts are about four thousand miles from the center of motion.

PUPIL. I suppose then you infer that as the centrifugal force is greater the farther it is removed from the center, that the parts near the poles have a tendency to fly off towards the equator.

TUTOR. I do. And as we have finished this part of our subject, I shall take leave of you.

> 15. Here it must be observed that there will be a little variation from sun-rising to sun-setting, as the earth is advancing in its orbit.

DIALOGUE XI.

TUTOR.

I now propose giving you a description of the moon, and I doubt not it will afford you some degree of pleasure.

PUPIL. Indeed it will, as I know little more than that she is a secondary planet or satellite, revolving round the earth, and with it round the sun.

TUTOR. You know her mean distance from the earth.

PUPIL. I did not recollect that: 240 thousand miles.

TUTOR. Right. Her diameter is about 2161 miles, and her bulk about a fiftieth part of the earth's. Her axis is almost perpendicular to the plane of the ecliptic, consequently she can have no diversity of seasons.

PUPIL. What is her period?

TUTOR. The time she takes to revolve from one point of the heavens to the same again is called her *siderial* or *periodical revolution*, and is performed in 27 days, 7 hours, 43 minutes; but *synodical revolution*, or the time taken up to revolve from the sun to the same apparent situation with respect to the sun again, or from change to change, is 29 days, 12 hours, and 44 minutes.

PUPIL. I do not clearly comprehend it.

TUTOR. If the earth had no annual motion, the period of the moon would be uniformly 27 days, 7 hours, 43 minutes; but you are to consider that whilst the moon is revolving round the earth, the earth is advancing in its orbit, and of course she must be so much longer in completing her synodical revolution as the difference of time between that and her siderial revolution. This I will make clear to you in a few minutes.—What is the situation of the hour-hand and minute-hand of a watch at twelve o'clock?

PUPIL. They will be in conjunction.

TUTOR. And will they be in conjunction at one?

PUPIL. No, Sir.

TUTOR. Yet the minute-hand has made a complete revolution: but before they can be in conjunction again the minute-hand must move forward till it overtakes the hour-hand.

PUPIL. I now understand it, and must beg you to explain to me the different phases of the moon.

TUTOR. Take this ivory ball, and suspend it by the string with your hand between your eye and the candle. Let the candle represent the sun, the ball the moon, and your head the earth. In this situation, as the candle enlightens only one half of the ball, the part turned from you will be enlightened, and the part turned to you will be dark. This will be a representation of the moon at change, and as no part of her enlightened hemisphere is turned to the earth, she can reflect no light upon it, and consequently is invisible to us. She now rises and sets nearly with the sun.—Turn yourself a little to the left, and you will observe a streak of light like what is called the new moon.

PUPIL. I see it clearly.

TUTOR. Move round one quarter.

PUPIL. One half of the side next me is now enlightened.

TUTOR. You may conceive it to be the moon at first quarter.—Go on, and you will see the light increase till the ball is opposite to the candle, when the side next you will be wholly illumined, and will give you a just idea of the moon at full, which now rises about the time of sun-setting, being opposite to the sun: and, the farther she advances in her orbit the later she rises.

PUPIL. It is plain it must be so. She rises with the sun at change, being then in conjunction: and as she revolves in her orbit the same way as the earth does on its axis, the earth will have farther to revolve each day before it can see the moon. At the full she is in opposition, and of course rises when the sun sets: and so continues to rise later and later, till the change again.

TUTOR. You imagine that the moon rises exactly with the sun when she is at change; and when he sets, at full. I will presently convince you of your mistake; and would have you now proceed with your ball. Place it again opposite to the candle, and as you turn round you will find the light gradually decrease as it before increased, that the side that was before enlightened is now dark, and the dark side light. When you have gone three quarters round, one half of the side next you will be enlightened, and will resemble the moon at last quarter. As you go on the darkened part will increase, till you arrive at the place you set off from, where the light is quite obscured.

PUPIL. I have now completed the circuit, and am much delighted with it, as by this simple contrivance I can perceive the various changes of the moon, and that the western side is enlightened from the change to the full, and the eastern side from the full to the change.

TUTOR. I find then it has fully answered the purpose intended.

PUPIL. Indeed it has. But if you will give me leave I will use the ball again.

TUTOR. By all means.

PUPIL. I perceive, as I move round, that the same side of the ball is turned towards me whilst every part is turned to the candle. Is it so with the moon?

TUTOR. It is: and as every part of the moon is turned to the sun, she makes one revolution on her axis whilst she makes one in her orbit.

PUPIL. This is very singular. If the same side of the moon be always turned to the earth, the opposite side of course can never see it.

TUTOR. And they must likewise be deprived of the earth as a moon.

PUPIL. True. But how is it known that the same side of the moon is always opposed to the earth?

TUTOR. The moon, like our earth, consists of mountains and valleys, which, when seen through a good telescope, are very beautiful. The mountainous parts appear as lucid spots and bright streaks of light: and as the same spots, &c. are constantly turned to the earth, she must keep the same side to the earth.

PUPIL. It is very clear. Are there no seas?

TUTOR. It was formerly imagined that the dark parts were seas, but later observations prove that they are hollow places or caverns, which do not reflect the light of the sun. Besides, if there were seas there would consequently be exhalations, and if exhalations, clouds and vapours, and an atmosphere to support them. That there are no clouds is evident, because when our atmosphere is clear, and the moon above our horizon in the night-time, all her parts appear constantly with the same clear, serene, and calm aspect.

PUPIL. Has the moon then no atmosphere?

TUTOR. If she has it is imperceptible to us: for, when she approaches any star, we cannot discover with our best telescopes any change of colour or diminution of lustre in the star till the instant it is lost behind her: whence it is clear, that she can have no such gross medium as our atmosphere to surround her.

PUPIL. May we not then doubt whether she be inhabited or not, as without air we cannot breathe?

TUTOR. The same Almighty Being who created us and gave us air to breathe, may have provided a different way for their existence. It does not hold good that, because we could not live there, she is not inhabited. Fish will live a considerable time in water under an exhausted receiver: and, I have heard of a toad being found in a block of marble. Your doubt therefore, I think, ought not to be admitted.

PUPIL. I am satisfied. And must now beg to be informed how I may observe the moon's motion.

TUTOR. Her real motion round the earth, may be easily known by remarking when she is near any particular star. Thus, suppose you see her west, that is to the right of it, she will be approaching, then in conjunction with, and afterwards pass it towards the east. Her apparent motion is that of rising and setting, which is occasioned by the rotation of the earth on its axis.

PUPIL. I remember not long since, when you shewed me Jupiter, that the moon was west of him: the next evening I saw her almost appear to touch him, and soon after at a great distance from him easterly. I now see that her real motion is from west by south to east, and her apparent motion from east by south to west.

TUTOR. If you have no objection, I will now explain the cause of eclipses.

PUPIL. So far from it, that it will give me the greatest pleasure.

TUTOR. Take your ivory ball, suspend it as before, in a right line between your eye and the candle.—Can you see the candle?

PUPIL. No, Sir.

TUTOR. For what reason.

PUPIL. Because the ball prevents the light coming to me.

TUTOR. This then represents an eclipse of the sun, which can never happen but when the moon is between the sun and the earth, which must be at the change: for, as light passes in a right line, the sun is hidden to that part of the earth which is under the moon, and therefore he must be eclipsed. If the whole of the sun be obscured by the body of the moon, the eclipse is total: if only a part be darkened, it is a partial eclipse; and so many twelfth parts of the sun's diameter, as the moon covers, so many digits are said to be eclipsed.

PUPIL. May not the word digit be applied to the moon as well as the sun?

TUTOR. It may: for it means a twelfth part of the diameter of either the sun, or the moon.

PUPIL. As you have now shewn me the cause of an eclipse of the sun, I am anxious to have that of the moon explained.

TUTOR. We must again have recourse to your little ball.—Turn yourself round till it is opposite to the candle in a line with your head, and you will see that no light can be thrown on it from the candle, because your head is between them. In like manner the rays of the sun are prevented falling on the moon, by the interposition of the earth: she must therefore be eclipsed.

PUPIL. I see it clearly. And as an eclipse of the sun happens when the moon is at change, that of the moon must be when she is at full; for, it is then only the earth's shadow can fall on the moon, the earth being at no other time between the sun and her.

TUTOR. The diameter of the shadow is about three times that of the moon, and consequently the moon must be totally eclipsed whilst she continues in it. On the contrary, the shadow of the moon at an eclipse of the sun, covers so small a part of the earth's surface, that the sun is totally or centrally eclipsed to but a small part of it; and its duration is very short. But a faint or partial shadow surrounds this darkened shade, in which the sun is more or less eclipsed, as the place is nearer to or farther from its center; this partial shadow is called the *penumbra*. I have prepared for you a little drawing, representing an eclipse both of the sun and moon, which I think will enable you better to understand what I have been explaining. (Plate IV. Fig. 1 and 2.) In the former, *p. p.* is the penumbra.

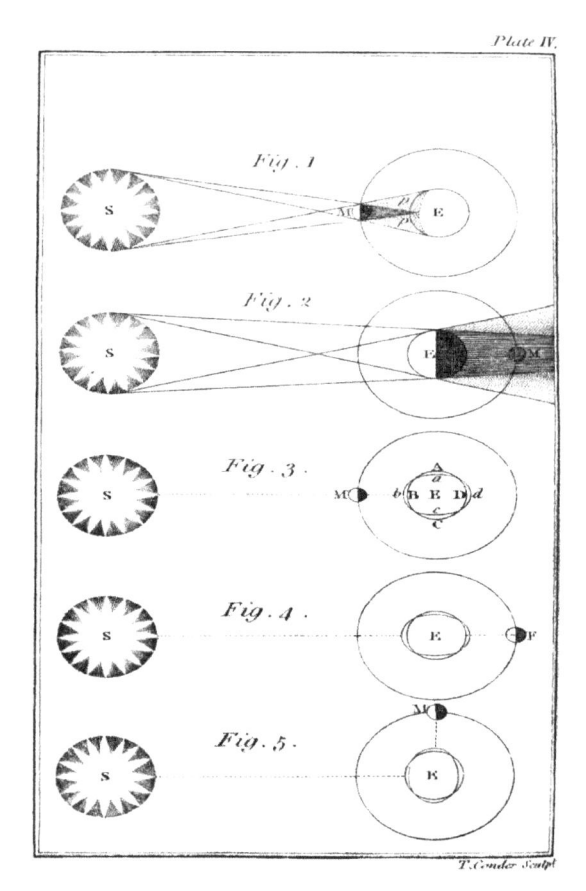

PUPIL. In what does a central differ from a total eclipse?

TUTOR. An eclipse of the sun may be central, and not total; for, those who are under the point of the dark shadow, will see the edge of the sun like a fine luminous ring, all around the dark body of the moon when the sun is eclipsed at the moon's greatest distance from the earth; but when she is nearest the earth at an eclipse of the sun, the eclipse is total. When the penumbra first touches the earth, the general eclipse begins; when it leaves the earth, the general eclipse ends. An eclipse of the moon always begins on the moon's eastern side, and goes off on her western side; but an eclipse of the sun begins on the sun's western side, and goes off on his eastern side. When the moon is eclipsed in either of her nodes, the eclipse is both central and total.

PUPIL. Pray, what is the reason we have not an eclipse at every full and change of the moon?

TUTOR. For the same reason that Mercury and Venus are not seen to pass over she sun's disc at every inferior conjunction.

PUPIL. Is the orbit of the moon then inclined to the plane of the ecliptic?

TUTOR. It is: and no eclipse of the sun can happen but when the moon is within 17 degrees of either of her nodes: neither can there be one of the moon, unless she be within 12 degrees. At all other new moons she passeth either above or below the sun, as seen from the earth: and at all other full moons above or below the earth's shadow, according as she is north or south of the ecliptic. You now see that the moon must sometimes rise before and sometimes after the sun at change, and before or after he sets at full.

PUPIL. I do, Sir, and am much obliged to you for this pleasing account of the moon, and of eclipses: and if you have any thing farther to observe, it will afford me additional pleasure.

TUTOR. You may, at some time or other, have an opportunity of seeing a total eclipse of the moon; it will therefore be necessary to prepare you for a phænomenon which otherwise you might be much surprized at, and that is, that after the moon is immersed in the earth's shadow, she is still visible.

PUPIL. This is a phænomenon that I am not able to account for; for, the moon being an opaque body, she cannot shine by her own light[16], and the rays of the sun are prevented falling on her by the interposition of the earth, she cannot therefore shine by reflection.

TUTOR. It is by reflection that we see her; for the rays of the sun which fall upon our atmosphere are refracted or bent into the earth's shadow, and so falling upon the moon are reflected back to us. If we had no atmosphere, she would be totally dark, and of course invisible to us.

PUPIL. What is her appearance?

TUTOR. It is that of a dusky colour, somewhat like tarnished copper.—I have one thing more to remark before we quit this subject, which is, that the moon's nodes have a retrograde or backward motion, in a direction contrary to the earth's annual motion, and go through all the signs and degrees of the ecliptic in little less than nineteen years, when there will be a regular period of eclipses, or return of the same eclipses for many ages.

PUPIL. Pray, Sir, what do you propose for our next subject?

TUTOR. The ebbing and flowing of the sea, or cause of the tides.

16. Dr. Herschell supposes the moon and the rest of the planets may have some inherent light: the side of the planet Venus, turned from the sun, having been seen, as we see the moon soon after the change.

DIALOGUE XII.

TUTOR.

In order to explain the cause of the tides, I have since I saw you last prepared a little drawing for you, (Plate IV. fig. 3.) where S represents the sun, M the moon at change, E the center of the earth, and A B C D its surface, covered with water. It is obvious, from the principles of gravitation, that if the earth were at rest the water in the ocean would be truly spherical, if its figure were not altered by the action of some other power. But, daily experience proves that it is continually agitated.

PUPIL. What is the cause of this agitation?

TUTOR. The attraction of the sun and moon, particularly the latter: for, as she is so much nearer the earth than the sun, she attracts with a much greater force than he does, and consequently raises the water much higher, which, being a fluid, loses as it were its gravitating power, and yields to their superior force.

PUPIL. What proportion does the attractive power of the sun bear to that of the moon?

TUTOR. As three to ten. So when the moon is at change, the sun and moon being in conjunction, or on the same side of the earth, the action of both bodies is on the surface of the water, the moon raising it ten parts,[17] and the sun three, the sum of which is thirteen parts, represented by B *b*. Now it is evident, that if thirteen parts be added by the attractive power of those bodies, the same number of parts must be drawn off from some other part, as A *a*, C *c*. It will now be high-water under the moon at *b*, and its opposite side *d*, and low-water at *a* and *c*.

PUPIL. That the attraction of the sun and moon must occasion a swelling of the waters on the side next them, I can readily conceive, and that this swell must cause a falling off at the sides: but that the tide should rise as high on the side opposite to the sun and moon, in a direction contrary to their attraction, is what I am not able to account for.

TUTOR. This difficulty will be removed when you consider that all bodies moving in circles have a constant tendency to fly off from their centers. Now, as the earth and moon move round their center of gravity, that part of the earth which is at any time opposite to the moon will have a greater centrifugal force than the side next her, and at the earth's center the centrifugal force exactly balances the attractive force: therefore, as much water is thrown off by the centrifugal force on the side opposite to the moon, as is raised on the side next her by her attraction. Hence, it is plain, that at D, fig. 3, the centrifugal force must be greater than at the center E, and at E than B, because the part D is farther from the center of motion than the part B. On the contrary, the part B being nearer the moon than the center E, the attracting power must there be strongest, and weakest at D. And, as the two opposing powers balance each other at the earth's center, the tides will rise as high on that side from the moon, by the excess of the centrifugal force, as they rise on the side next her by the excess of her attraction.

PUPIL. In this explanation you have mentioned nothing of the sun.

TUTOR. From what I have already said it must be plain to you that if there were no moon the sun by his attraction would raise a small tide on the side next him; and, it is as evident that the tides opposite would be raised as high by the centrifugal force: for the sun and earth, as well as the earth and moon, move round their center of gravity. This may be exemplified by an easy experiment. Take a flexible hoop, suppose of thin brass, tie a string to it and whirl it round your head, and it will assume an elliptical shape; the tightness of the string drawing out the side next to your hand, and the centrifugal force throwing off the other.

PUPIL. This I clearly comprehend.

TUTOR. I shall now refer you to the next figure, (fig. 4.) where F represents the moon at full: the sun and moon are in opposition, and yet the tide is as high on each side as in the former case. I wish you to shew me the cause.

PUPIL. I will use my endeavour to do it, Sir.

TUTOR. Then I doubt not you will accomplish it.

PUPIL. When the moon is at full, ten parts of water are raised from that side of the earth next her, by her attraction; and, as the side which is next her is opposite to the sun, three parts must be thrown off by his centrifugal force, the sum of which will be thirteen parts next the moon.—From the side opposite to the moon, and under the sun, ten parts are thrown off by her centrifugal force, and three raised by his attraction, making thirteen, the same as before.

TUTOR. I could not have done it better. These are called *Spring Tides*. But when the moon is in her quarters, the action of the sun and moon are in opposition to each other; that is, they act in contrary directions (see fig. 5.) The moon of herself would raise the water ten parts under her, and throw off ten parts by her centrifugal force on the opposite side; but, the sun being then in a line with the low-water, his action keeps the tides from falling so low there, and consequently from rising so high under and opposite to her. His power, therefore, on the low-water being three parts, leaves only seven parts for the high water, under and oppose the moon. These are called *Neap Tides*.

PUPIL. This is very plain.

TUTOR. You would naturally suppose that the tides ought to be highest directly under and opposite to the moon: that is, when the moon is due north and south. But we find, that in open seas, where the water flows freely, the moon is generally past the north and south meridian when it is high-water. For, if the moon's attraction were to cease when she was past the meridian, the motion of ascent communicated to the water before that time would make it continue to rise for some time after: as the heat of the day is greater at three o'clock in the afternoon than it is at twelve; and it is hotter in July and August than in June, when the sun is highest and the days are longest.

PUPIL. These are convincing reasons. And, pray what time after the moon has passed the meridian, is it high-water?

TUTOR. If the earth were entirely covered with water, so that the tides might regularly follow the moon, she would always be three hours past the meridian of any given place when the tide was at the highest at that place. But, as the earth is not covered with water, the tides do not always answer to the same distance of the moon from the meridian at the same places, because the regular course of the tides is much interrupted by the different capes and corners of the land running out into the oceans and seas in different directions, and also by their running through shoals and channels. But, at whatever distance the moon is from the meridian on any given day, at any place, when the tide is at its height there, it will be so again the next day, much about the time when the moon is at the like distance from the meridian again.

PUPIL. Are not the tides later every day than they were the preceding day?

TUTOR. Yes; and the reason is obvious: for, whilst the earth is revolving on its axis in twenty-four hours, the moon will be advancing in her orbit; therefore the earth must turn as much more than round its axis before the same place which was under her can come to the same place again with respect to her, as she has advanced in her orbit during that interval of time, which is 50 minutes. This being divided by 4, gives 12-1/2 minutes; so that it will be 6 hours 12-1/2 minutes from high to low-water, and the same time from low to high-water: or 12 hours 25 minutes from high-water to high-water again.

PUPIL. This I understand perfectly well.

TUTOR. I have now finished my description of the tides, and having a little time to spare, if you wish to know how to find the proportionate magnitude of the planets with that of the earth, and to calculate their distances from the sun, I will employ it that way.

PUPIL. At our first conference I remember you shewed me the proportion that the other planets bear to the earth, with their periods and distances from the sun; but to have it in my power to make the calculations myself, will certainly give me great pleasure.

TUTOR. To find what proportion any planet bears to the earth; or, that one globe bears to another, you must observe that, *all spheres or globes are in proportion to one another as the cubes of their diameters*. So that you have nothing more to do than to cube the diameter of each, and divide the greatest by the least number, and the quotient will shew you the proportion that one bears to the other.

PUPIL. The operation appears very simple; but, as I do not know what a cube number is, I cannot perform it.

TUTOR. You cannot forget what a square number is.

PUPIL. The product of any number multiplied into itself is a square number, as 4 is the square of 2.

TUTOR. Any square number multiplied by its root, or first power, will be a cube number. Thus 4 multiplied by 2 will be 8, which is the cube of 2; 9 is the square or second power, and 27 the cube or third power of 3, &c. This you will perhaps better understand by

A TABLE OF

Roots.	1.	2.	3.	4.	5.	6.	7.	8.	9.
Squares.	1.	4.	9.	16.	25.	36.	49.	64.	81.
Cubes.	1.	8.	27.	64.	125.	216.	343.	512.	729.

PUPIL. I do, Sir; and am now prepared for an example.

TUTOR. The diameter of the sun is 893552 miles, of the earth 7920 miles; how much does the sun exceed the earth in magnitude?

PUPIL. The cube of 893522, the sun's diameter, is 713371492260872648; and of 7920, the earth's, 496793088000. And 713371492260872648 divided by 496793088000 is equal to 1435952, and so many times is the bulk of the sun greater than that of the earth.

TUTOR. This one example may suffice, as I intend by and by to give you a table of diameters, &c.; you may then calculate the rest at your leisure.

PUPIL. I shall now, Sir, be glad to have the other explained.

TUTOR. The periods of the planets, or the times they take to complete their revolutions in their orbits, are exactly known; and the mean distance of the earth from the sun has been also ascertained. Here, then, we have the periods of all, and the mean distance of one, to find the distances of the rest; which may be found by attending to the following proportion:

As the square of the period of any one planet,

Is to the cube of its mean distance from the sun;

So is the square of the period of any other planet,

To the cube of its mean distance.

The cube root of this quotient will be the distance sought.

PUPIL. Here again I find myself at a loss, as I have not learnt to extract the cube root.

TUTOR. I will give you [118]Doctor Turner's rule, which I think will answer your purpose.

"First, having set down the given number, or resolvend, make a dot over the unit figure, and so on over every third figure (towards the left hand in whole numbers, but towards the right hand in decimals); and so many dots as there are, so many figures will be in the root.

Next, seek the nearest cube to the first period; place its root in the quotient, and its cube set under the first period. Subtract it therefrom; and to the remainder bring down one figure only of the next period, which will be a dividend.

Then, square the figure put in the quotient, and multiply it by 3, for a divisor. Seek how often this divisor may be had in the dividend, and set the figure in the quotient, which will be the second place in the root.

Now, cube the figures in the root, and subtract it from the two first periods of the resolvend; and to the remainder bring down the first figure of the next period, for a new dividend. Square the figures in the quotient, and multiply it by 3, for a new divisor; then proceed in all respects as before, till the whole is finished."

The following example will, I trust, make it clear to you.

EXAMPLE.

It is required to find the cube root of 15625.

$$
\begin{array}{r}
\overset{..}{}\\
15625 \quad\quad (25 \\
8 \\
\hline
12)76 \\
15625 \\
\hline
\cdots\cdot
\end{array}
$$

Point every third figure, and the first period will be 15; the nearest cube to which, in the table I gave you just now, you will find to be 8, and its root 2; the 8 you must place under the 15, and the 2 in the quotient: take 8 from 15 and 7 will remain, to which bring down 6, the first figure of the next period, and you have 76 for a dividend. The figure put in the quotient is 2, the square of which is 4, which multiplied by 3 is 12, for a divisor. Now 12 in 76 will be 5 times; cube 25, and you will have 15625, which, subtract from the resolvend, and nothing will remain; which shews that the resolvend is a cube number, and 25 its root.

PUPIL. You say 12 in 76 is 5 times; I should have said 6 times.

TUTOR. In common division it would be so; but as the cube of 26 would be greater than the resolvend from which you are to subtract it, it can go but 5 times.

PUPIL. Now, Sir, I think I have a sufficient knowledge of the rule to solve a problem.

TUTOR. The earth's period is 365 days, and its mean distance from the sun 95 millions of miles; the period of Mercury is 88 days—what is his mean distance?

PUPIL. As the distance of the earth is given, I must make the square of 365 the first term, the cube of 95 the second, and the square of 88 the third term of the proportion.

TUTOR. Certainly.—Take your slate, or a piece of paper, prepare your numbers, and make your proportion.

PUPIL. I find the square of 365 = 133225; of 88 = 7744; and the cube of 95 = 857375.

Then 133225 : 857375 :: 7744 to a fourth term.

I now multiply the second and third terms together, and divide the product by the first, the quotient 49836 is the cube of the mean distance of Mercury from the sun in millions of miles, and the fourth term sought.

TUTOR. So far you are right. Now extract the root.

	49836	(36		3
	27			3
27)	228		Sq. of 3 =	9
	46656		Mul. by	3
	3180		Divisor	27

Cube of 36 =

PUPIL. The root I find to be 36, which is the mean distance of Mercury from the sun, in millions of miles.

TUTOR. You now see, that although 27 in 228 will go 8 times, yet here it will go but 6 times; and, as there is a remainder, it shews you that the resolvend is not a cube number.

PUPIL. I see it clearly.

TUTOR. You now seem perfect in the rule; I shall therefore not trouble you with any more examples, but shall give you the table I promised you.

TABLE.

Names of the PLANETS.	Diameters, in English Miles.	Magnitude, compared with the Earth.	Periods, in Years and Days.	Mean Distance from the Sun, in Mil. of Miles.

Sun	[A]893522	1435952	—	—
Mercury	3261	1/14	0 — 88	36
Venus	7699	5/49	0 — 224	68
Earth	7920	1	1 or 365	95
Moon	2161	1/49	—	—
Mars	5312	1/3	1 and 322	145
Jupiter	90255	1479	11 — 314	494
Saturn	80012	1031	29 — 167	906
Georgian	34217	82	83 — 121	1812

A. The Diameters were taken from Adams's Lectures, Vol. IV. p. 39.

PUPIL. I shall take the first opportunity of calculating the rest, in which I am certain I shall have great satisfaction.

TUTOR. I have now conducted you through the elementary parts of astronomy, have given you a general view of the system of the world, and prepared you to pursue the study with profit and pleasure.—In your future researches, the more accurate you are, the more you will discover of regularity, symmetry, and order in the constitution of the frame of nature.

"Hail, Sov'reign Goodness! all-productive Mind!

"On all thy works thyself inscrib'd we find;

"How various all, how variously endow'd,

"How great their number, and each part how good!

"How perfect then must the Great Parent shine, ⎫

"Who, with one act of energy divine, ⎬

"Laid the vast plan, and finish'd the design!" ⎭

THE END.